SYSTEMS AND MODELS FOR ENERGY AND ENVIRONMENTAL ANALYSIS

Systems and Models for Energy and Environmental Analysis

edited by

T. R. LAKSHMANAN
Boston University, U.S.A.

PETER NIJKAMP
Free University, Amsterdam

Gower

Published by
Gower Publishing Company Limited,
Gower House, Croft Road, Aldershot, Hampshire GU11 3HR,
England

British Library Cataloguing in Publication Data

Lakshmanan, T.R.
Systems and models for energy and environmental analysis.
1. Power resources - Mathematical models
I. Title II. Nijkamp, Peter
33.79'0724 HD9540.1

ISBN 0-566-00558 1

Printed and bound in Great Britain by
Biddles Ltd, Guildford and King's Lynn

Contents

1 Introduction

T. R. LAKSHMANAN AND PETER NIJKAMP

1. INTRODUCTION

In the last decade and a half, affluent societies have become increasingly concerned with new kinds of scarcities that appear to threaten their high quality of life. Ironically, these new scarcities stem from the very economic and technological progress that underlies their affluence.

First, there is the decline in the quality of the natural and built environment -- shortfalls in the quality of air, water and land, and congestion in public facilities and services -- brought about by the increasing production and consumption activities. Second is the rapid pace of depletion of energy and certain non-energy materials that has raised fears about the adequacy of materials to support future production and (high amenity) consumption. Further, there has been concern about the distributional consequences of these environmental and energy and non-energy materials scarcities. The unequal distribution of the burden of environmental deterioration, and the unbalanced supply of energy and non-energy resources have begun to pose political conflicts. Indeed, these environmental and material scarcities and the associated equity problems demonstrate a clear conflictual trend within regions in a country and between countries. Illustrations of such conflicts are provided by the current discussions on the working of a market system versus a planning system, on maximizing growth versus assuring environmental quality, and on the aim of assured energy supplies versus political and ecological risks of high energy production.

A key analytical problem relevant to these issues is the question of how to alleviate these scarcities while maintaining equitable economic progress and quality of life, in a situation which increasingly looks like a steady-state economy. Such harmonizing planning strategies for diversified social systems require methods for the resolution of conflicts arising from the interdependence between various parts of the systems. Integrated policy analysis aims at providing such tools for harmonizing goals and conflicts among individuals and for harmonizing among groups in society. Such integrated policy methods, developed jointly by social scientists and engineers, should attempt to identify feasible states of a social system such that these states reflect a meaningful compromise among different "pure" policy options.

The development of integrated policy analysis is aided by two distinct but related lines of inquiry. First, there is a clear need to articulate and elaborate on the multiple dimensions of the interactions between economic, energy, and environmental matters. As a complex description of the system emerges, integrated frameworks for economic-environmental-energy analysis need to be specified. Section II of this book provides a few examples of such integrated frameworks that provide disciplined delineations of the rich interactions among energy, environment and the economy.

Second, there is an urgent need to design and implement systems and policy models, consistent with the current state-of-the-art, to help assess a variety of policy proposals being tried to alleviate environmental and energy scarcities in Western Europe and North America. Section III provides four examples of such integrated energy-environmental-economic policy models in the Dutch system. Illustrations of such policy modelling in the U.S. appear in Section IV.

Peter Nijkamp introduces the readers in Chapter 2 to the multidimensionality of conflicts inherent in the economic-environmental-energy interactions. He proposes a multidimensional measure of welfare consistent with an integrated framework of economic-environmental energy analysis. A multidimensional interactive policy analysis framework based on multiple criteria analysis is proposed and illustrated for a case of choice between economic growth and energy conservation. Finally, Nijkamp sets forth a broad range of currently available multidimensional models, organized according to their applicability to a) the continuous, or discrete cases or b) the availability of hard (quantitative, metric) or soft (ordinal, qualitative) information.

Nijkamp and Rietveld focus in Chapter 3 on the spatial differences in the economic, environmental and energy structures that promote policy frictions and uncoordinated decisions among regions. They propose a multilevel multiobjective definition of a spatial system, that would promote coordination between policy issues and between different sectors or regions at different levels so as to increase overall efficiency and equity. They present the analytical methods for the solution of such problems, identifying their operationality and flexibility of application to a broad range of environmental and energy problems.

Thomas Wilbanks explores in Chapter 4 the many-sided implications of energy self-sufficiency option pursued seriously for some time after the oil embargo in 1973 in the U.S. He examines the realism and welfare implications of some of these strategies for regional autarchy in energy extraction and use. He notes that complete self-sufficiency is somewhat an illusory goal when there is often so much regional interdependence with regard to other factors of production. Energy self-determination, which may make some sense from the point of view of participatory goals, can also mean considerable cost, unless mechanisms for free capital flow or for conflict resolution are worked out. Wilbanks makes a strong plea for further analysis of the issues he has raised regarding what energy self-sufficiency would mean for regions, countries and for the international political economy.

Antzen and Braat present in Section III an integrated environmental model that is designed for the assessment of economic, social, and ecological effects of an urban development plan proposed for a polder region in Southwest Netherlands. The noteworthy features of this model are (a) the effort to integrate four processes (normally handled in isolation) -- demographic, economic, ecological and private and public facility provision, (b) the use and integration of existing qualitative and quantitative data and (c) the regional scale of analysis. This model is a clear illustration of the emerging integrated policy modelling approaches where contributions are being jointly made by social scientists, natural scientists, and engineers in a context where there are considerable costs of communication among experts trained in different scientific "cultures".

Lesuis, Muller and Nijkamp present an operational integrated interregional input-output model for Netherlands. This model focuses on the spatial aspects of integrated energy-economic-environmental interactions, so that policy conflicts within and between regions can be analyzed. Two innovative attributes of the model are the use of a translog price frontier to incorporate input substitution and technology change in the model and the formulation of a generalized multiobjective programming framework to analyze the interdependencies among various elements of the policy structure.

Hafkamp and Nijkamp present a regional model of income, labor opportunities, environmental quality, and, indirectly, of energy use. Specifically, three interrelated models -- a national-regional economic model, a regional sectoral employment model, and a pollution model against a background of energy availability -- are presented in detail. Work is underway on this ambitious model which can make a significant contribution to comprehensive analysis given its economic and process structures and scenario methods for decision analysis.

Bannink, Broekhof and Nijkamp present a programming approach to the design of economic development policy. A dynamic input-output model, (extended to include capacity limits) reformulated as a programming model, is broadened to include pollution emissions and energy consumption. Since a programming model formulated here cannot guarantee an equilibium with the outside world, some relationships reflecting the balance of trade and the public sector are also specified. The stage is then set for defining alternative objective functions relating to income, employment, etc., and for solving for alternative futures.

Hordijk et al. provide the most disaggregated analysis of the relationships between the production pattern, pollution and energy consumption to date in the Netherlands -- 60 economic sectors, 17 types of energy and 55 pollutants. Input-output analysis permits estimation of total pollution, and total energy consumption per unit of final product and the "balance of pollution" (as a counterpart to the balance of trade). A number of 1985 scenarios of structural abatement of pollution constructed by the use of linear programming, with constraints on environmental and economic targets, are compared. The authors conclude that, notwithstanding strong doubts whether in the long run continued economic growth is compatible with an acceptable environmental quality, there are opportunities currently to contribute to the amelioration of the pollution problem.

Ratick and Lakshmanan present in Section IV an overview of a large, comprehensive modelling system, termed the Strategic Environmental Assessment System (SEAS), that has been extensively used for assessment of environmental and energy policies in the U.S. SEAS is viewed as a second generation model evolving out of the strategic concerns in the last decade and a half with the long term capacity of the physical world to provide for our expanding needs for natural resources and for a high quality environment. SEAS is an integrated series of computer implemented models and data bases concerned with the economy, energy and environment and organized in a framework for conditional analysis of how economic and demographic activities affect and are reliant upon the physical environment. Ratick and Lakshmanan provide brief descriptions of the various SEAS modules -- a collection of economic, engineering and process descriptions of the environment, energy and the economy. While the 'transparency' of SEAS allows it to be used for many types of applications and in many different ways, its complexity and flexibility limit its application to teams of analysts and systems specialists. A number of representative applications of SEAS in U.S. Environmental Protection Agency and U.S. Department of Energy and other national and regional agencies are identified illustrating the information available from SEAS in one such typical application. Since SEAS is one of the more used policy models in this field, three such applications of SEAS are described in detail in the following chapters.

Dossani, Watson and Weygardt utilize SEAS as a primary tool to assess the economic and environmental consequences of alternative scenarios of solar energy development to year 2000 in the U.S. The three scenarios are termed as the Base Case (that assumes the solar provisions of National Energy Act and continuation of present trends), the maximum practical (base case augmented by comprehensive and aggressive solar initiatives) and technical limits (solar energy growth constrained only by productive capacity) scenarios. It appears that in the latter two high solar scenarios, front-end costs of capacity build-up to year 2000 exceed environmental and economic benefits. However, from the perspective of year 2000, if future benefits are discounted at 6%, the maximum practical scenario would be an acceptable project. Since this scenario generates substantial external environmental benefits ($175 billion by year 2000), which will not be generated in the "free market" base case scenario, the higher solar energy levels are societally more appropriate. Though overall environmental and economic benefits are greater in the higher solar scenarios, their sectoral and regional incidence varies by scenario. A consideration of these is vital if solar energy implementation strategies are to be implemented. Finally, Dossani, Watson and Weygandt caution us that their overall assessment of the net benefits of the maximum practical scenario may be affected by the issues not considered -- e.g., the savings rate required for solar investment and environmental effects.

The application of SEAS to an assessment of alternative transportation energy conservation scenarios is the theme of the paper by Patterson and the SEAS system is elaborated by providing greater detail in the transportation sector to form the TECNET System. Patterson uses the TECNET System to analyze six alternative transportation scenarios in the U.S. It is seen that indirect energy saving of conservation strategies do not necessarily follow the same pattern as the direct energy savings. For instance, the largest direct energy savings for automobiles comes from a scenario that causes an increase in indirect energy use.

Bennett and Reuss describe a strategic application of SEAS in the U.S. Environmental Protection Agency (EPA). EPA produces "Environmental Outlook Reports" that provide historical trends information on environmental matters, and future trends for a number of pollutants, along with the implications of such trends. Projections obtained from SEAS support this report. Another application of SEAS is as the major analytical tool in integrated environmental assessments of large scale development of energy resources (e.g., Western Coal Development, Ohio River Basin Development, etc.). Bennett and Reuss discuss the strengths and weaknesses of SEAS as they emerge from these applications. They suggest that a streamlined version of SEAS may be required to provide quick strategic assessments of emerging environmental problems in the future when fewer analytical resources are likely to be available.

David Greene explores another component - serious gasolene supply shortages - of the energy security problem that has concerned decision makers in the affluent oil importing countries since 1973. From a national perspective any strategy to manage supply shortages must address the twin issues of social costs of petroleum shortages (e.g., time and fuel wastes in gasolene lines and administrative costs) and the potential redistribution of income between consumers and owners of gasolene resources. Greene points out that regional variability in gasolene demand elasticities may further ameliorate or exacerbate the equity problem. He explores the implications of these variable elasticities on the various pricing and rationing schemes that have been proposed for managing petroleum shortages. While pricing strategies with mechanisms for taxing and transfering windfall profits appear to be relatively efficient and equitable, Greene is not sure if the petroleum and product transport system in the U.S. has the ability to carry out the necessary transshipment of supplies during such a shortage.

As we conclude this summary of large complex integrated models of the economy, energy and environment, an important point about this new genre of models needs to be made. The integrated models reported in Section III and IV above represent refinements and synthesis of disparate models drawn from component areas of interest. The latter include accounting models, econometric models, physical process models and programming models. The integration of these models imposes enormous demands on the model builder-analyst's intuitive and rational faculities. Starting at the interface between reality and the model, the analyst filters in the questions that are in the legitimate domain of the integrated model. He also assesses the model output, being on guard against anomalous results that signal error and interpreting results consistent with reality. Such judgments require the analyst -- or the team of analysts in the model such as SEAS -- to be intimately familiar with the inner workings of all parts of the model components, some of which come from different conceptual frameworks.

Given the size and complexity of these models, such capabilities currently reside largely in the model builder-analysts -- a situation not conducive for gaining the confidence and reliance of the broader policy making commmunity, in these models. Models such as SEAS which had to operate in the public policy arena from the beginning have documented the models and source data in detail and diffused the model knowledge and judgment capabilities to a team of modeller-policy analysts. However, if such models are to gain acceptance broadly in the policy community, there must be a cadre of professional model analysts, who can engage in sensitivity studies, probing shaky assumptions, identifying critical points, tracing through policy conclusions and understanding the simulated effects of policy choices. The contributions of such model analysts in studying and reporting on models would greatly stimulate detailed open documentation of all such models and data and promote the intellectual ties between model building and policy research. Such a development would greatly improve the policy relevance of integrated models in general and increase the chances of informed assessment of energy-environmental policies.

In light of the above-mentioned remarks, and given the recent experience in many countries, a research agenda for the analysis of integrated energy-environmental-economic development patterns should include the following items:

- A study of the impact (realized or expected) of changes in the energy sector (for example, price increases, decline in oil supply) on the spatial distribution of activities and vice versa. The sensitivity of location and settlement patterns as well as of their associated transportation - and mobility patterns deserves a closer analysis. The impacts of shifts in land use and in urban space (for example, the design of energy saving cities and transportation networks or the creation of satellite cities) upon energy consumption are fairly unknown as well.

- A thorough investigation of the differences in regional energy efficiency. Such a comparative study requires a closer look at the physical conditions of the regions concerned, the differences in technological conditions (different kinds of energy sources, possibilities of interfuel substitution, etc.) and the sectoral composition of production and final demand of the regional system at hand. Only an integrated analysis of these determinants may explain interregional energy efficiencies.

- An analysis of feasible regional policy and decision areas. In this respect, the use of policy sceanarios (for example, alternative strategies of covering the need for energy resources, alternative solutions for tackling sudden shocks in oil supply, etc.) may be a meaningful vehicle. This can also be combined with simulation models for exogenous international development.

- A study of the interactions between regional production, energy consumption, pollution and spatial allocation of activities.

- In this case a multidimensional approach is necessary to assess the trade-offs among conflicting items. Such a multidimensional policy analysis is a prerequisite for arriving at a balanced selective development, in which economic options and regional objectives are brought into harmony with sound ecological principles of environmental management. In this way, one may also obtain an integrated view of the impacts of oil price increases, of input substitution and of alternative technologies. Sudden shocks (for example, catastrophe theoretical types of perturbations) can be studied in this framework.

- A careful examination of the distributional impacts of energy problems for the spatial system at hand. Equity problems may emerge, among others through shifts in regional accessibilities as a consequence of energy saving physical planning through different spatial impacts of energy policies (for example, spatial differences in the rate structure of electricity) or through incapability of some regions to pay the higher energy costs. There is a fair chance that the energy problem will worsen the equity situation in detriment to lagging regions.

- The construction of an accurate and up-to-date information system on environmental impacts of production, consumption, mobility and energy use. In this respect, spatially disaggregated models are a prerequisite for arriving at balanced policy decisions aiming at cohenernt and effective solutions.

- The construction of integrated land use -environmental quality models in order to incorporate physical planning models in environmental models. In this, environmental impact statements, technological impact statements and urban impact statements may become meaningful vehicles. Parallel to this development attempt should also be made for linking such models and statements to policy evaluation methods. The recently developed multi-criteria models open many perspectives for a better integration of diverse components and interests in our complex society.

It has to been admitted that the successive items of this agenda cannot immediately be realized in an adequate manner. Many research tasks need to be undertaken in order to reach a mature analysis of integrated energy-environmental-economic policy problems.

Frameworks for Integrated Energy Environmental Analysis

2 Multidimensional Analyses of Economic-Environmental-Energy Problems: A Survey

PETER NIJKAMP

1. INTRODUCTION

During the seventies, the attention of social scientists, engineers, economists and planners has been increasingly focussed on environmental issues (pollution, decline in the quality of life, exhaustion of energy and other natural resources, etc.). Instead of the traditional scarcity of commodities, a new kind of scarcity appeares to affect the well-being in our affluent society: lack of clean air and fresh water, lack of energy and raw materials, lack of parks and recreation facilities, etc.

Most Western countries appear to show a conflictual trend: the material growth gives rise to a new scarcity which reduces or even neutralizes the initial economic and technological progress. The origins of this 'law of conservation of disaster' can be found in the first and second law of thermodynamics.

The first law is the law of conservation of matter and energy: in a closed system the total amount of material and energy is constant. Consequently, energy cannot be destroyed: only qualitative transformations are possible. This is also reflected in the materials balance model, which teaches us that any rise in throughput (e.g., consumption) will affect the input side through a decrease in resources, and it also will affect the output side through an increase in waste and pollution.

The second law explains the entropy principle: in a closed system the entropy (i.e., a measure for the amount of unavailable energy) tends to be at a maximum. This implies that, in the long run, all energy maybe transformed into heat which will be dispersed all over the system so that it is no longer usable. Clearly, without additional low entropy from outside (such as solar energy), energy cannot be recycled. Even recycling of materials may be difficult, because this also requires more low entropy energy. In conclusion, given a finite energy stock on earth, it is extremely important to seek alternatives which may guarantee a balanced supply of flow energy (mainly solar energy).

In the meantime, however, the industrialized world is increasingly confronted with the consequences of the new scarcity. Energy resources are no longer limitlessly available, at least not for low prices. The supply of energy inputs and of raw materials is overloaded with risks and uncertainties. Clearly, the precise stock of natural resources is unknown and it may even be true that - in the near future - the world as a whole will not run out of its natural resources, but a major problem is, of course, the unequal distribution of these resources, and the consequent political frictions. Another problem is the risk of some alternative energy resources (such as nuclear energy).

Consequently, the new scarcity has three aspects in the energy field: - physical: the ongoing exhaustion of energy resources may hamper a balanced development in the long run. The trade-offs between the need of physical inputs (i.e., energy), the satisfaction from the throughput (i.e., final demand) and the dissatisfaction from the environmental repercussions may give rise to sharp conflicts.

- economic: the increasing scarcity of energy resources also leads to price increases which may disturb a balanced economic development. Even though the energy prices in many countries do not reflect the real scarcity and political risks, it is clear that sudden and unanticipated price shocks may lead to perturbations in our economic system.

- political: The unequal geographical distribution of energy resources may lead to many political conflicts: oil is power! The power of a limited number of oil-producing countries may even lead to a purposely created, fictitious scarcity.

Given all these frictions, conservation of natural resources as well as more technological innovations may be regarded as wise responses to the intertwined energy-environmental-economic problems. A major analytical problem, however, concerns the question as to how to reconcile all these different interests in a situation which increasingly looks like a steady-state economy. In the rest of this paper an analytical framework will be sketched which is useful to deal with such problems in an operational sense.

2. A MULTIDIMENSIONAL FRAMEWORK OF ANALYSIS

The search for alternative energy resources and the stimuli needed to overcome the present economic stagnation lead, among others, to conflicts with environmental management. The new scarcity seems to be a trap for all kinds of future activities we are going to undertake. For example, in the past, several electricity plants in the Netherlands were using natural gas. Given the exhaustion of the stocks of natural gas, some of these plants have had to switch again to oil. This leads indeed to a conservation of natural gas reserves, but also increases the environmental burden due to the rise in sulfurdioxide. Thus, any kind of substitution effect in the energy sector must be dealt with carefully.

The central question is whether a reconciliation of divergent options is possible in an integrated energy-environmental-economic policy framework. It is evident that the resolution of problems related to the scarcity requires a new way of thinking about welfare of individuals, groups, cities, regions or nations.

The traditional way of measuring welfare is average income in one or the other form. This measure has been the major criterion for evaluating economic developments, welfare increases or growth perspectives. Recently, several authors have criticized this unidimensional welfare criterion for several reasons, among others, its neoclassical foundation, its neglect of intangibles, incommensurables and spillovers, its emphasis on quantitative aspects of welfare, etc. Clearly, this measure also does not reflect the importance to be attached to future generations, the risks and uncertainties in the supply of oil resources, and the non-depletable character of energy sources.

Consequently, average income cannot be used as a reliable measure for welfare. During the last decade the insight has grown that welfare is essentially a multidimensional variable which should comprise *inter-alia* average income, growth, environmental quality, distributional equity, availability of energy resources, spatial accessibility, and so forth. Therefore, the welfare of individuals, groups or regions has to be represented by a vector profile rather than by a scalar measure. In such a way, a more appropriate picture of welfare constituents can be obtained, as such a vector profile may also include unpriced consequences of human activities. Analogously, spatial distributional impacts can be taken into account.

The relevance of such a multidimensional measure can be demonstrated by means of the following simple example. Assume 2 countries with the same average income. Country 1 has no energy resources at all (Japan, e.g.), whereas country 2 is able to cover the major part of its energy needs from its own sources. Moreover, country 1 is highly polluted and densely populated, whereas country 2 has a pleasant environment and an agreeable quality of life. Could we now say that the welfare level of country 1 is equal to that of country 2? Most people will rightly argue that country 1 is far behind country 2. This implies already that income *per capita* is not a reliable welfare measure.

It should be noticed that the use of profile methods for representing welfare is not entirely new. This multidimensional view of welfare has its roots in the social indicator movement, followed among others by the environmental impact analyses. There are, however, two basic shortcomings in these traditional impact analyses. First, there is a lack of coherence among the variables constituting the welfare components. The first shortcoming can be overcome by making a systematic subdivision of the welfare constituents according to major welfare categories. For example, a certain welfare profile $\underline{w}$ might be subdivided into the following main categories: economic, social, physical-geographical, environmental and energy categories. For a meaningful analysis of such a wide variety of welfare constituents it may next be appropriate to make a further subdivision into subprofiles:

$\underline{w}^E$:	economic:	average income
		growth
		investments
		consumption
		differentiation of economic structure
		specialization index, etc.
$\underline{w}^S$:	social:	situation on the labor market
		degree of unionization
		degree of social stability

		social facilities medical health care, etc.
$\underline{w}^P$:	physical-geographical:	location pattern residential land use industrial land use volume of transport network population density congestion, etc.
$\underline{w}^M$:	environmental:	ecological quality indicators pollution quality of natural areas, etc.
$\underline{w}^N$:	energy:	energy use investments in energy technology degree of interfuel substitution international political risks, etc.

It is clear that such a multidimensional welfare profile can be constructed separately for each country, region or group. Clearly, this requires a lot of detailed information, although very often ordinal or even qualitative information may be satisfactory.

The second shortcoming in the use of many traditional impact analyses concerns the lack of coherence. Normally, there is an interdependency between the successive welfare constituents. For example, the use of energy is related to the average production and income, etc. Therefore, it is, in general, necessary to take account of such interdependencies by means of a formal model describing the structure of the successive relationships. Basic ingredients of such a model are: production, investments, income, consumption, pollution, employment and energy consumption. A simple illustration of such a formal model is given in Figure 1.

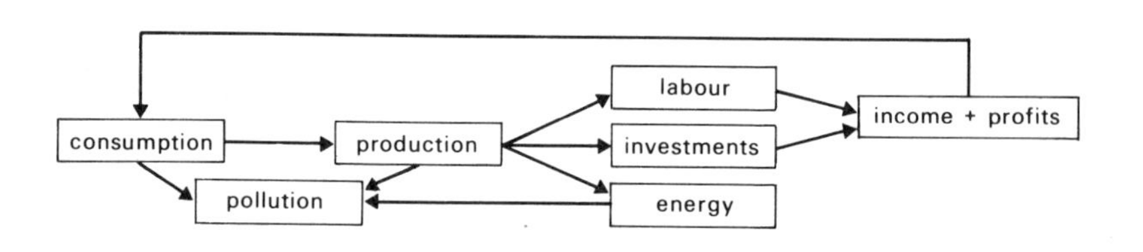

Fig. 1. A simplified economic-environmental-energy picture.

A significant advantage of a multidimensional approach is that many aspects of human welfare falling outside the realm of the traditional price mechanism (such as externalities) can be taken into account, so that a fairly complete picture of human welfare profiles can be obtained.

The foregoing multidimensional framework will also be used in this paper as a frame of reference for analyzing conflicts in the energy sector. The next section will be devoted to a further exposition of this approach in a choice and policy setting.

3. MULTIDIMENSIONAL CHOICE PROBLEMS IN THE ENERGY SECTOR

A multidimensional utility framework precludes the application of simple optimality rules for decisions in the energy sector: incommensurable goals, conflicts between policy objectives or between policy-makers and intangible or external impacts hamper a traditional unidimensional approach.

The conflicting nature of economic-environmental-energy policy decisions can be illustrated by means of Figures 2 and 3, which present an illustration of the trajectory of economic growth and the available energy resources, respectively. These figures clearly demonstrate the conflicts between diverging options: more economic growth means less energy resources! Therefore, the problem of finding a balanced growth path is of great importance. This requires, however, an operational framework for reconciling conflicting objectives.

In this respect, a multidimensional policy analysis may be extremely important in order to find a compromise for many dilemmas. The formal framework of multidimensional policy analysis is based on multiobjective programming or multiple criteria analysis.

Let us assume that a decision-maker or a decision-committee has to make a choice between two completely conflicting options, viz. economic growth and energy conservation. It is clear that normally a choice in favor of economic growth will lead to a reduction of energy conservation and _visa versa_. The degree to which these issues are conflicting can be identified on the basis of a formal model which describes all structural relationships between the variables concerned (see Fig. 1) (including technical, political, social or economic constraints on certain variables). Such a conflict between different objectives can be represented by means of a so-called efficiency frontier (see Fig. 4). This efficiency frontier represents the set of feasible and efficient solutions of the one variable for any given value of the other one, and _visa versa_. This curve is an attainment frontier which represents all feasible optimal combinations of economic growth and energy conservation. This curve indicates that no further increase in economic growth can be attained without a decrease in energy conservation, and _visa versa_ In principle, all points on this curve can be identified through a parametric variation of the two tradeoffs.

It is clear that any good solution should be a point on the efficiency locus. The curve AB is essentially the only relevant part, because all other points outside this part are dominated by the edge AB. Of course, the precise choice of a specific point on the curve AB is the result of political priorities and a political decision process.

There are evidently many ways to arrive at a compromise point on the edge AB. In the framework of multiobjective programming models and multiple criteria analysis a wide variety of methods and techniques has been developed which aim at identifying a certain compromise between divergent objectives. In the next section, one possible solution

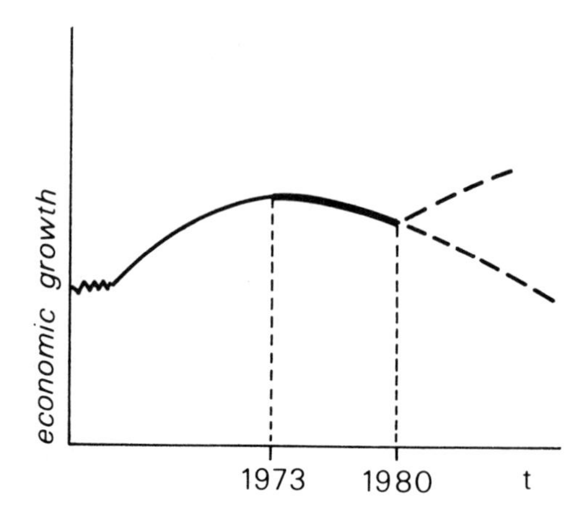

Figure 2. An Illustrative Trajectory for Economic Growth.

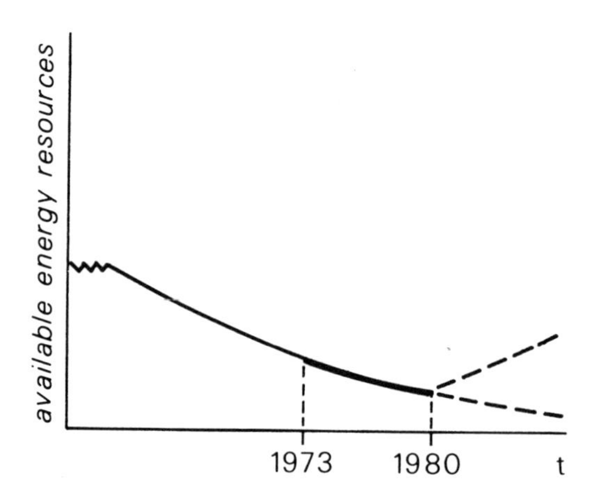

Figure 3. An Illustrative trajectory for the Available Energy Resources.

method, viz. the interactive method between the analyst and the policy-maker(s) will be explained in greater detail.

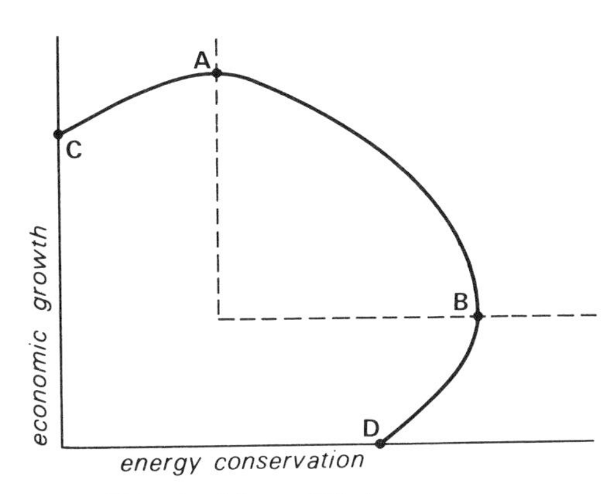

Figure 4. An illustrative efficiency curve.

4. AN INTERACTIVE POLICY FRAMEWORK

An interactive policy analysis is based on an exchange of information between the policy-maker(s) and the analyst(s). Especially, in the framework of process planning in which decisions are not taken once at a time but require several stages during a certain time period, interactive analyses may be extremely useful.

The basic idea of an interactive method is that the 'satisficer' principle provides a useful paradigm to find, after a number of stages, an ultimate compromise solution. Normally, the following steps are undertaken: the analyst suggests a certain initial (trial) solution to the policy-maker, the policy-maker indicates whether or not he is satisfied with this solution, the information from the policy-maker is used as a constraint in the next stage of the analysis in which the analyst suggests an adjusted solution, and so forth. In Fig. 5 the interactive mechanism is illustrated in greater detail.

Suppose we have identified the maximum level of economic growth and of energy conservation. This solution is called the ideal point. It is clear that this point is not feasible, but it serves as a frame of reference for finding a first compromise on the efficiency curve. This compromise can be found by minimizing the distance between the ideal point and the efficiency focus.

It should be noted that a compromise point might also be found by maximizing the distance to the origin (which is the worst situation). Alternative methods (for example, min-max strategies) might be applied as well.

After the identification of this first compromise solution, the decision-maker is asked whether or not he is satisfied with the trial solution. If not, he has to indicate which values of the successive objective functions are unsatisfactory for him. Next, this information is included as a side-condition in the next run of the model, so that the whole procedure can be repeated, until finally a converging solution is attained (see Fig. 6).

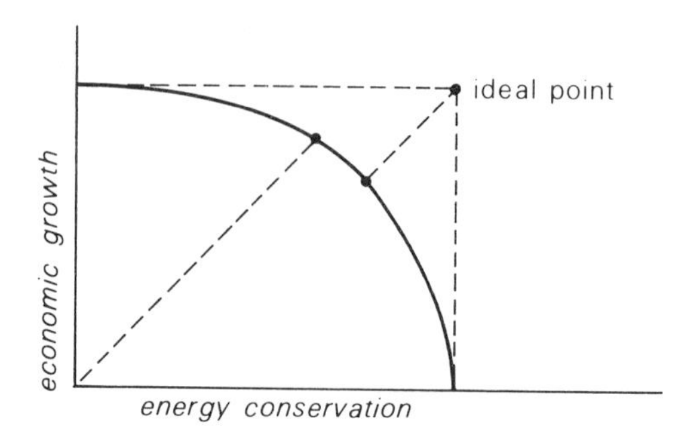

Fig. 5. The efficiency-frontier of an ideal point approach.

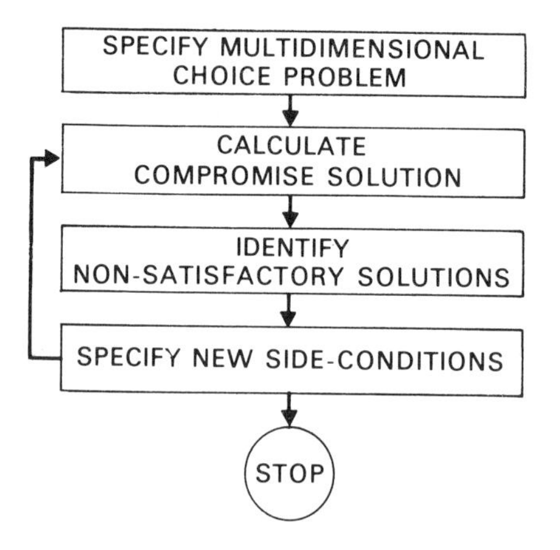

Fig. 6. Steps of an interactive multidimensional choice problem.

One of the advantages of the latter interactive approach is that it leads to a satisfactory compromise solution without using, necessarily, information on weights or preferences. Instead, the information gathered during the successive stages of the policy process is used in order to find an ultimate compromise solution. Another advantage is that these approaches lead to an active role for both decision-makers and analysts during a process which provides enough information to identify a reasonable compromise.

5. CLASSES OF MULTIDIMENSIONAL ENERGY POLICY PROBLEMS

The foregoing introduction to multidimensional choice problems was rather general and none of the multidimensional models have been specified. There are, however, a wide variety of multi-dimensional models for energy policy analysis. A feature of all these models is the existence of multiple policy criteria or objectives which determine the ultimate decision.

A first class concerns continuous energy models. For example, how many barrels of oil should be imported in order to satisfy the energy needs, the environmental constraints and the economic growth targets.

A second class is related to discrete energy plans. For example, how many electricity plants have to be built and where should they be located? The latter types of problems are integer decision problems related to distinct plans or projects (see Fig. 7).

In addition to the abovementioned subdivision, one may also subdivide multidimensional models according to the availability of hard (quantitative, metric) information or soft (ordinal, qualitative) information.

		INFORMATION	
		HARD	SOFT
DISCRETE	MULTI-CRITERIA MODELS	I	III
CONTINUOUS	MULTI-OBJECTIVE MODELS	II	IV

Figure 7. Classes of multi-dimensional models.

Models of type I are among others:

- trade-off method
- expected value method
- correspondence analysis
- entropy analysis
- discrepancy analysis
- concordance analysis
- goals - achievement method

Models of type II are among others:

- utility models
- penalty models
- constraint models
- goal-programming models
- hierarchical models
- min-max models
- ideal point models

Soft models of type III are among others:

- expected value method
- lexicographic method
- ordinal concordance method
- permutation method
- metagame analysis
- eigenvalue method
- frequency method
- multi-dimensional sealing method

Finally, models of type IV are among others:

- fuzzy set models
- stochastic models
- soft econometric models

The majority of these models can easily be used in the interactive framework explained before. This implies that in the area of integrated energy-economic-environmental policy models a wide variety of theories, methods and techniques does exist which may support more balanced policy decisions.

3 Multilevel Multiregional and Multiobjective Policy Models for Environmental and Energy Management

P. NIJKAMP AND P. RIETVELD

1. INTRODUCTION

Environmental and energy policy analysis is a mode of thinking in a field which is full of conflicts and dilemmas. Examples of such conflicting issues are: the working of the price and market system *versus* a more centralized or planned system, the aim of a maximum production growth versus environmental interests, the aim of a sufficient energy supply *versus* risks and ecological disturbances, a central policy co-ordination versus a regional decentralization, etc.

The present paper aims at providing a framework for analysing such conflicts. Its emphasis is (1) on multiobjective programming as a tool for identifying compromise solutions in conflicting decision problems and (2) on multilevel programming as a tool for attaining a satisfactory co-ordination between different decision levels (characterized by specific interests) of a policy system. This paper attempts to give an introduction to both modes of thinking, followed by a synthesis. A survey and classification of various kinds of multiobjective and multilevel decision problems will be given as well, while also the relevance of such approaches for environmental and energy policy analysis will be indicated.

2. CONFLICTS IN ENVIRONMENTAL AND ENERGY POLICY ANALYSIS

As mentioned above, environmental and energy management may lead to various conflictive policies and developments. As almost all regions in developed countries demonstrated a rapid growth in the use of energy resources, they found themselves confronted with greater external dependency. In addition, these regions were faced with mounting levels of waste and pollution arising out of the consumption and production of commmodities. Consequently, particularly the technologically advanced regions have to orient their consumption and production activities to a conservation of energy resources and a preservation of environmental quality.

One may expect that the current problems of high-consumption technological societies will evoke an adjustment process (either through the market system or by public intervention) toward a different consumption and production pattern. Examples of such adjustment processes are: the production (and consumption) of products with a greater durability, the design of a less energy-intensive technology, the construction of more energy-efficient power plants, or the creation of an adjusted location, settlement and transportation system.

It is clear that any change in the composition of the regional production and consumption structure or in the spatial lay-out of a society will have impacts on the regional development patterns. The same holds true for any change in the energy prices and in the supply of energy resources. It is plausible that such impacts will be greater as the changes in technological, economic or political circumstances will be more discontinuous in nature. The risks and uncertainties in the present oil supply are certainly a real danger for an unbalanced regional development pattern. In addition to the price system, alternative policy instruments should not be neglected (for example, rationing, standard setting, etc.); the choice of adequate policy instruments is one of the dimensions of a conflictive policy problem.

There is another conflicting element involved in a policy analysis of energy resources. Several energy activities pose serious threats to human health and to the environment. Any reduction in such threats may induce higher energy costs. The existence of unacceptable threats may even preclude a further development of new energy resources (as is demonstrated by the discussion about nuclear energy plants). Thus, energy policy analysis has to be put in a broader multidimensional framework, in which conservation of energy (including interfuel substitution), preservation of environmental quality and maintenance of a reasonable welfare level are simultaneously taken into account.

The energy situation has also many important spatial aspects. Due to differences in regional production technologies, the regional sectoral energy coefficients may show much variation. The same holds true for energy use arising from final consumption. However, even the different states of technology among regions do not explain entirely that the regional disparities in energy consumption per capita may be strikingly wide (even within the same country). A major reason for the occurrence of substantial regional disparities in energy consumption is the difference in sectoral composition of the regions. Another reason for the occurrence of disparities in energy use among regions may be - apart from differences in climate and physical conditions - a difference in the spatial location and settlement patterns of these regions.

Such spatial differences in the economic, environmental and energy structure lead obviously to several policy frictions among regions of a spatial system, and decisions of the one region may be neutralized by opposite, uncoordinated decisions in other regions. Therefore, a framework for arriving at compromises between different policy issues and between different actors (regions, e.g.) at different levels may be necessary to increase social efficiency and equity.

In general, a multilevel, multiobjective structure of a spatial system with different actors may be represented as follows (arrows denote conflicting interests).

In order to elaborate on Fig. 1, in the following sections more explicit attention will be paid to multiobjective and multilevel programming problems.

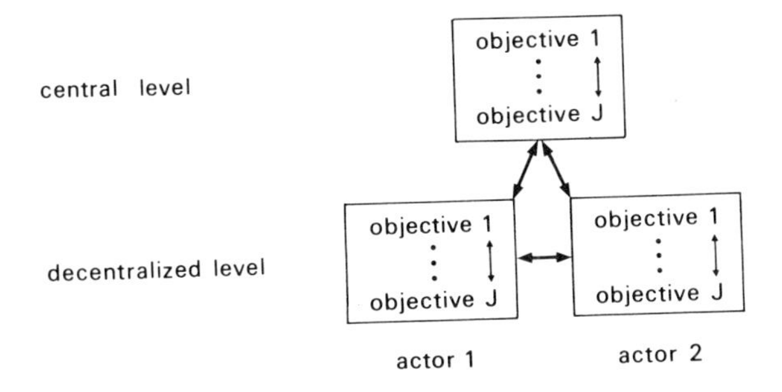

Fig. 1. A Multiobjective Multilevel policy framework.

3. MULTIOBJECTIVE PROGRAMMING

Multiobjective programming methods form a rapidly expanding field of research in operations research and management science. For some recent surveys we refer to Cohon [1978], Hwang and Masud [1979], Nijkamp [1979] and Rietveld [1980]. The aim of these methods is the analysis and solution of decision problems in which a decision-maker (DM) (or several DMs) faces several conflicting objectives.

The methods are devoted to such activities as:

- the analysis of the conflicts inherent in a decision problem, for example, by determining a series of alternatives in which a best result is strived after for only one objective so as to study the consequences for the other objectives.

- the generation of a representative set of efficient (Pareto-optimal) alternatives.

- the determination of alternatives reflecting a certain compromise between conflicting points of view.

- modelling preference statements of DMs to find alternatives which are in agreement with the DM's priorities.

- detailed analyses of the pros and cons of alternatives in order to rank the alternatives in order of attractiveness - given the DM's priorities.

It is an essential feature of multiobjective programming methods, that an exchange of information takes place between an analyst and a DM. The analyst generates information about the structure of the decision problem by analyzing conflicts and compromises. The DM informs the analyst about his preferences, as far as he is able to express them.

In <u>interactive</u> multiobjective decision methods this information exchange is structured as a process consisting of several runs. The basic idea of interactive methods is that first by means of a standard rule a provisional compromise solution is calculated by the analyst which has to be judged by the DM. Then the DM has to indicate which of the proposed compromise values of the objectives are not satisfactory to him. These preference statements can be incorporated by the analyst as side-conditions in the next run of the analysis. Then the procedure may be repeated again and again, until finally a converging satisfactory compromise solution has been identified.

4. MULTILEVEL PROGRAMMING

Multilevel programming models provide a framework for the coordination of decisions made in the various components of a system. The aim of multilevel programming is the achievement of an efficient distribution of resources and liabilities (e.g., energy, manpower) among components and an efficient management of spillovers between components (e.g., spatial diffusion of pollution).

It is assumed that the coordination has to be accomplished by a central unit which has the political power to give certain directives to the components. The central policy unit has only fragmentary knowledge about the structure of the decision problem. Hence, in addition to the communication process described in the preceding section, another communication process has to be introduced, aiming at providing the central unit with sufficient information about the structure of the decision problem.

The information exchange between centre and components takes place by means of essentially two kinds of variables:

1) quantities

2) (shadow) prices or productivities.

In *direct* methods, the central unit informs each component about the quantity of the various common resources allocated to this component. Then the components report the productivities of these resources back to the centre. Given this information about productivities, the central unit generates a new distribution of common resources which is again proposed to the components. It can be shown that under certain conditions such an information exchange converges to an efficient allocation of resources (cf. Ten Kate [1972] and Johansen [1978].

Indirect methods consist of a similar interaction, the difference being that now the centre confronts the components with prices of common resources, and the components report back the quantities demanded of these resources. Dantzig [1963] had provided a proof of convergence for these methods.

5. A CLASSIFICATION OF MULTIOBJECTIVE MULTILEVEL PROGRAMMING METHODS

In this section we will present a classification of integrated multiobjective and multilevel programming methods. In Table 1, these methods have been classified according to four dimensions:

(a) The *components* the centre interacts with:

- regions $r = 1, \ldots, R$

- policy sectors $j = 1, \ldots, J$. (e.g., economic development, employment, environment, energy)

TABLE 1. CLASSES OF MULTIOBJECTIVE MULTILEVEL PROGRAMMING METHODS

Components interacting with center	r = 1, ..., R			j = 1, ..., J			(j,r) = (1,1), (1,2), ..., (J,R)	
Objective of component	ω_r	ω_r	$\sum_j \lambda_{jr}\omega_{jr}$	ω_j	ω_j	$\sum_r \mu_{jr}\omega_{jr}$	ω_{jr}	ω_{jr}
Objective of centre	$\sum_r \omega_r$	$\Sigma\gamma_r\omega_r$	$\sum_{jr} \gamma_{jr}\omega_{jr}$	$\sum_j \omega_j$	$\sum_j \gamma_j \omega_j$	$\sum_{jr} \gamma_{jr}\omega_{jr}$	$\sum_{jr} \omega_{jr}$	$\sum_{jr} \gamma_{jr}\omega_{jr}$
Direct method — known weights		2	4		7	9		12
Direct method	1			6			11	
Direct method — not known weights		3	5		8	10		13
Indirect method — known weights		15	17		20	22		25
Indirect method	14			19			24	
Indirect method — not known weights		16	18		21	23		26

- regional policy sectors $(j,r) = (1,1), (1,2), \ldots, (1,R), (2,1), \ldots, (J,R)$.

For the ease of presentation, l will be used as a general indicator of a component ($l = 1, \ldots, L$). Consequently, $l = r, j, (j,r)$ for the three cases mentioned above.

(b) The structure of the objective functions of the centre (U_c) and the components (U_l) and the relationships between them. The following combinations can be distinguished:

- U_l is one-dimensional and U_c is the unweighed sum of the U_l (e.g., national energy consumption is the sum of regional energy consumptions).

- U_l is one-dimensional and U_c is the weighed sum of the U_l (e.g., the national objective function is the sum of the outcomes for a number of policy sectors, each supplied with an appropriate weight).

- U_l and U_c are multidimensional. The weights attached to the various objectives by the centre and the components are not necessarily identical.

(c) The type of information flowing upward and downward (see Section 4 for the distinction between direct and indirect methods).

(d) The availability of information about the weights to be attached to the various objectives. In Section 3 it has been indicated that in the case of "no information available" interactive multiobjective programming may be helpful to determine satisfactory solutions.

We will use the following notations to sketch the various classes of multiobjective multilevel programming methods. Let $\underline{x}_l$ denote the vector of decision and state variables of component l. In multilevel programming, there are essentially two types of constraints:

- constraints holding for each component separately, e.g.,

$$B_l \underline{x}_l \leq \underline{b}_l \ , \ \underline{x}_l \geq \underline{0} \ , \qquad (1)$$

where B_l and $\underline{b}_l$ denote a matrix and vectors of appropriate size.

- constraints holding for several components simultaneously, implying interdependence of components, e.g.,

$$\sum_l A_l \underline{x}_l = \underline{a} \ . \qquad (2)$$

The vector $\underline{a}$ denotes the available amounts of the common resources.

Given this information, the central decision problem reads:

$$\begin{aligned} \max! \quad & U_C(\underline{x}_1, \ldots, \underline{x}_L) \\ \text{s.t.} \quad & \sum_l A_l \underline{x}_l \leq \underline{a} \\ & B_l \underline{x}_l \leq b_l \qquad l = 1, \ldots, L \\ & \underline{x}_l \geq 0 \qquad l = 1, \ldots, L \end{aligned} \tag{3}$$

Since we assume that the centre has incomplete information on the B_l and $\underline{b}_l$, decentralization of decisions is necessary. In direct methods of decentralization, the decision problem of component l reads:

$$\begin{aligned} \max! \quad & U_l(\underline{x}_l) \\ \text{s.t.} \quad & A_l \underline{x}_l \leq \underline{a}_l \\ & B_l \underline{x}_l \leq \underline{b}_l \\ & x_l \geq 0 \end{aligned} \tag{4}$$

where $\underline{a}_l$ is the amount of the common resources assigned to component l ($\Sigma \underline{a}_l = \underline{a}$). The indices reported back to the centre are the marginal productivities in the use of $\underline{a}_l$ with respect to U_l. These indices can be determined as the shadow prices of the constraints $A_l \underline{x}_l \leq \underline{a}_l$.

In indirect methods, the decision problems of the components read:

$$\begin{aligned} \max! \quad & U_l(\underline{x}_l) - c(A_l \underline{x}_l) \\ \text{s.t.} \quad & B_l \underline{x}_l \leq \underline{b}_l \\ & \underline{x}_l \geq \underline{0} \end{aligned} \tag{5}$$

where $c(A_l \underline{x}_l)$ denotes the costs, charged by the centre because of the use of common resources. In this case, the indices reported back to the centre are the quantities $A_l \underline{x}_l$ of the common resources demanded by the component concerned.

In this contribution we will not treat in detail how the centre generates the indices for the components. Suffice it to say that by an appropriate reformulation of (3), the centre can digest past responses of components to generate new indices leading ultimately to convergence (cf. Dantzig [1963], Ten Kate [1972], and Nijkamp and Rietveld [1980].

We will now give a more detailed description of the indices exchanged in some of the classes of methods distinguished in Table 1.

Class 1. The indices produced by the centre are quantities $\underline{a}_r$ for each region r. The indices reported back are the marginal productivities of the resources $\underline{a}_r$ with respect to U_r. Class 6 and 11 are similar to 1. From the view point of the indices exchanged, also the classes 2, 7 and 12 are similar to 1, the only difference being that the centre weights the productivities reported by means of certain factors .

Class 14. The centre submits prices π to charge the regions for their use of scarce resources. These prices hold for all regions. The regions report back the quantities demanded of the common resources. The classes 15, 19, 20 and 24 are similar to class 14.

Class 4. The effect of the multidimensional character of the objectives is that the productivities reported to the centre also have to be multidimensional. Thus, for each common resource k, the region r reports the marginal productivity with respect to all individual objectives ω_{jr}. Consequently, when K is the number of common resources, each region generates J.K indices in virtue of the centre. An important consequence of divergent weights is, that the components may use the resources in a way, different from the intentions of the centre. Thus, from the centre's view point sub-optimal solutions may be reached. Class 9 is similar to class 4.

Class 17. The multidimensionality of objectives has for this class consequences, which are comparable to class 4. For each common resource k, the centre generates a vector of prices $(\pi_{1kr}, \ldots, \pi_{Jkr})$ to be charged to the regions for their use of common resources. Consequently, the total number of indices produced by the centre equals J. K. R. Similar classes can be found in cells 22 and 25 of Table 1.

Class 5. The introduction of uncertainty about weights means for the interactive multilevel methods that a second process of interactions has to be built in, namely between DMs and analysts, at the level of the centre as well as of the components. The resulting process can be sketched as follows:

(a) Interaction at the central level between the central DM and his analysts to determine a satisfactory provisional distribution of common resources.

(b) This provisional distribution is proposed to the regions.

(c) Interaction at the level of the regions between each regional DM and his analysts to determine a satisfactory allocation of the resources.

(d) Each region reports back J . K shadow prices corresponding to this allocation (cf. class 4).

(e) Go back to (a) until convergence has been reached.

The classes 3, 8, 10 and 13 have the same structure as class 5. The classes 16, 18, 21, 23 and 26 can be found as a straightforward adaptation of the above-mentioned structure to the characteristics of indirect methods as shown in classes 14 and 17.

6. CONCLUSION

The above-mentioned multiregional, multilevel methods have a general scope, but also a particular relevance for environmental and energy policies, in so far as these policies are being performed at different levels and with conflicting priorities among actors (such as regions). Given the operational nature of these methods and the wide variety of such methods, they may be regarded as flexible tools for a large set of environmental and energy management problems.

BIBLIOGRAPHY

Dantzig, G., Linear Programming and Extensions, Princeton University Press, Princeton. (1963)

Johansen, L.,Lectures on Macroeconomic Planning, North Holland Publ. Co., Amsterdam. (1978).

Kate, A. ten, "Decomposition of Linear Programs by Direct Distribution", Econometrica, Vol. 40, pp. 883-898. (1972).

Nijkamp, P., Multidimensional Spatial Data and Decision Analysis, Wiley, New York. (1979).

Nijkamp, P. and P. Rietveld, "Multiobjective Multilevel Policy Models", European Economic Review, Vol. 15, pp. 63-89. (1981).

Rietveld, P. Multiobjective Decision Methods and Regional Planning, North Holland Publ. Co., Amsterdam. (1980).

Cohon, J. L., Multiobjective Programming and Planning, Academic Press, New York. (1978).

Hwang, C. L. and A.S.M. Masud, Multiple Objective Decision Making-Methods and Applications, Springer, Berlin. (1979).

4 Energy Self-Sufficiency as an Issue in Regional and National Development

THOMAS J. WILBANKS

When one's well-being is in the hands of others and their benevolence and constancy are in doubt, one tends to want to reduce that dependency. For this reason, regions and nations throughout the world are confronted with a felt need to become more self-sufficient in energy supply.

This paper will address the general question: is energy self-sufficiency good for regional and national development or is it bad. Because the ideas in the relevant literatures are fragmented and sometimes contradictory (and perhaps because paradoxes are a part of life), it is impossible at this point to provide a neat and rigorous answer; but a number of the key ideas will be reviewed and, based on them, some speculations will be offered about the connections between energy self-sufficiency and regional growth and development.

1. SELF-SUFFICIENCY AS AN ISSUE

A. U.S. National and Regional Perspectives

In the United States, public interest in energy self-sufficiency began with the 1973 oil embargo, which had a profound psychological impact on us. Before that, we had thought that economic linkages between nations and regions were generally a good thing, that linkages meant opportunity and efficiency and higher levels of well-being. During the 1960's, we worked hard to get the nations of the world to agree to reduce barriers to free trade. Two hundred years before, we had made this point of view one of the pillars of our Constitution, stipulating that there should be no barriers to interstate commerce. Still today, our textbooks tell students that accessibility is one of the most important of all comparative advantages.

In 1973, however, we discovered that linkages can mean vulnerability as well as opportunity; they can have adverse effects on well-being as well as positive ones. We know now that the health of our economy can depend on decisions that people in other countries make about energy supply levels, prices, and conditions of supply -- decisions over which we may have very little influence. And the resulting sense of impotence has become a powerful political force.

One consequence, of course, has been a search for ways to reduce our vulnerability. The first call was for a "Project Independence," to make the U.S. self-sufficient in energy supply by 1985. More recently, the federal government has considered restricting oil imports by quotas or tariffs, and it has suggested that we ought to be willing to pay a very high price to develop domestic energy alternatives that are not now competitive in the marketplace. For instance, the Synthetic Fuels Corporation was created in 1980 with an initial budget estimate of $88 billion, or about $1600 from every family of four in the entire country. A major study by the Office of Policy and Evaluation in the Department of Energy, conducted during the summer of 1980 as the foundation for National Energy Plan III, focused exclusively on reducing our dependency on oil imports by 1985 (Department of Energy, 1980). It is quite clear, therefore, that in the early 1980's the United States equates greater energy self-sufficiency with reduced economic and political vulnerability and that the country's leaders are convinced it is worth paying a price to get such a benefit.

The growing awareness of a relationship between dependence and vulnerability has also influenced the perspectives of regions and states within the United States. Presaged by the state of Oregon, which in 1972-73 conducted a wide-ranging evaluation of the suitability of energy supply technologies for Oregon (e.g., State of Oregon, 1973), this has led to investments by states such as California and New York in energy R&D to meet their own needs, filling gaps that they perceive in the profile of federal energy R&D. It has stimulated a major study to see if California could meet all of its future energy resources with indigenous energy resources (Lawrence Berkeley Laboratory, 1978).[1] It has led producing states to institute (or advocate) restraints on interstate trade, such as severance taxes or separate intrastate energy markets, either to raise revenues or to attract industry by offering abundant intrastate energy supplies. And the U.S. Congress has discussed measures related to state and regional energy self-sufficiency, such as requiring regional energy plans to be prepared, requiring states that generate radioactive wastes to dispose of their own, and involving states more actively in energy facility siting and permitting processes.[2] From the standpoint of states and regions, the benefits of increased self-sufficiency include not only comparative advantages for producing states and vulnerability reduction for importing states but such desirable qualities as participation and local control.

B. Less-Developed Country Perspectives

In many of the less-developed countries (LDC's), the concerns are even more acute than in the United States. Frequently, oil imports are critically important as a source of energy for industry, transportation, and even -- in the form of kerosene or other petroleum products -- for agriculture and individual households. As world oil prices have skyrocketed, many LDC's have had to borrow heavily to continue essential energy purchases, and many credit lines are virtually exhausted. For such countries, the time energy crisis is at hand <u>regardless</u> of what happens with exports from the Middle East.

A few examples will illustrate how desperate the situation is in many LDC's. In some countries, such as Cameroon, energy purchases now require up to 50 percent of household income (Cecelski, Dunkerley, and Ramsey, 1979; Mbi, 1980a). One effect, worldwide, is rapid deforestation as people seek cheaper energy sources near at hand, threatening long-term ecological damage on top of difficulties with credit limitations and balance of payments. Such impacts are especially dramatic when oil supplies are disrupted for any reason, which could happen if and when any LDC is unable to continue borrowing in order to compete with wealthier countries in the world oil market. For instance, the landlocked country of Burundi in Central Africa receives all of its oil by rail on a line through Uganda. During the war in Uganda, Burundi's oil supply was cut off completely. Although the interruption was relatively brief, a result was that now, within a 70-mile radius of the capital city, there is not a single tree standing (Mbi, 1980b).

It is not surprising that LDC's are seeking to use their own energy resources to get out of this bind. The problem is that the energy import situation they are trying to change is itself an impediment to bringing the changes about. As one example (Islam, 1980), the major balance of payments problem in Bangladesh is purchases of imported oil. A major consumer is industry in such centers as Chittagong. About 300 miles from Chittagong is a natural gas field with an estimated 50 year supply of gas; the price of the gas at the well-head is one-sixth the price (per equivalent energy unit) of the oil at the port. Bangladesh, however, finds it virtually impossible to borrow the funds needed to build a pipeline from the resource to the demand center. Normal credit lines have been exhausted buying oil. Credit from private sources is difficult to get, because the perceived instability of the economy and government (partly due to energy conditions) makes a contract too risky. Meanwhile, the current situation adds to the instability of the economy and government. It's a vicious circle.

In spite of challenges such as this one, most LDC's are doing their best to find ways to become energy self-sufficient -- or nearly so. This is a central objective of many requests for international assistance, and it is a major emphasis of country energy assessments (e.g., Department of Energy, 1979a and 1979b). A few, winners in the "resource lottery" (Haggett, 1975; 460) and blessed with adequate capital supplies, appear to be succeeding: Brazil, Argentina, Mexico. But for most, it remains a dream.

C. Self-sufficiency and Consensus

As an issue, self-sufficiency is closely related to the way decisions are made. Most people agree, for example, that our energy crisis in the United States is a social crisis, not a technological one. We do not lack technologies and domestic resources that can provide us the quantities of energy that we want, in forms that are familiar to us. We are capable of developing technologies to meet a wide range of conditions for resource use, environmental protection, and human safety. Yet, in the six years since the 1973-74 oil embargo, our dependence on oil imports has increased, as has the likelihood of serious and disruptive energy shortages in the near future. It is hard to interpret this as evidence that we are making progress in solving our energy problems.

The heart of the matter is our process of making decisions (Kash and others, 1976; Wilbanks, 1980). Clearly, we are having a great deal of trouble coming to agreement among ourselves whenever our decisions are going to involve social costs as well as social benefits (which is increasingly the case in energy actions). Once we reach a broad consensus among a wide range of participants in energy policymaking -- whatever we agree -- we can usually make it work. But without that consensus, we are unlikely to be able to make, or at least to sustain and implement, the specific decisions that enable us to meet our energy goals.

To some extent, dissensus has always been a part of the American scene, a corollary of our democratic ideals. But a fundamental kind of socio-political change in the past several decades has made it a more salient issue in energy policy. As recently as two decades ago, it was generally assumed that major decisions in the United States could be grouped into distinct categories, in each of which a limited number of groups had a right to participate -- usually those with direct economic, regulatory, or technical roles. For instance, it was quite clear who made oil policy decisions and utility policy decisions and national defense decisions. As long as the participants were agreed, an action could be taken. By the end of the 1950's, in fact, the ability of these decision-making consortia, often made up of big business and big government, was so unbridled that President Eisenhower felt it necessary to warn the country about the power of a "military-industrial complex."

But a number of important events during the 1960's, including civil rights struggles, Vietnam, and the Santa Barbara oil spill, convinced many people that decisions being made within the traditional frameworks were affecting individuals and groups outside those frameworks. As a result, the demand grew for broader participation in decisionmaking, ranging from pressures for consumer representation on corporate boards of directors to student participation in promotion decisions for university professors. The "environmental movement" was the most visible indicator of this change, but it involved more than environmental interests alone.

In contrast to the 1950's, energy policy decisions now involve a wide range of groups and interests as parties to the decisions, and it takes a broad consensus (or at least acquiescence) among the parties to take major actions (Kash and others, 1976; Schurr and others, 1979). This change is probably irreversible, and it is in many respects the way a democratic process is supposed to work. But it does leave us with a serious problem. As Lewis Branscomb (1978) has suggested, our decision-making structures -- designed for a different time and a different set of conditions -- have broken down under the current conditions, and we do not have a new structure to replace them. Without it, our old structures turn uncertainties into disagreements, and disagreements into antagonisms, and prospects for action fade away time after time.

Resolving this problem requires a fresh look at how consensual decision-making is most likely to take place. Both experience and theory indicate that the answer may be to localize decisions, in some sense. In other words, to become more energy self-sufficient as a nation, we may need to encourage smaller units within the nation to become more self-sufficient, at least in their energy decision-making.

For instance, as the demand for participation has grown in the past twenty years, people who were outside the traditional decision-making frameworks have found more often than not that their only channel for entering the process is through government: by voting for or against candidates, by getting legislators to take actions that would give people a way to use the courts or formal hearings to get involved, etc. And the government unit or jurisdiction to which most people have ready access is relatively small: a ward, a city, an election unit, at most a state. One result has been that the balance of power in our federal system has shifted, at least to some degree, away from functional subdivisions toward areal subdivisions -- in other words, units defined more by scale than by function, where a person's right to participate (in a sense, his or her share or ownership) is determined by citizenship rather than by wealth.

As another example of recent experience, the National Coal Policy Project, an important pioneering effort by environmentalists and industry executives to seek areas of agreement about the use of coal, concluded that energy supply facilities should be located in the vicinity of those who will be using the energy (Murray, ed., 1978). The people who benefit should bear the costs; they should be the ones to decide what mix of benefits and costs is best for them, operating at a scale where such decisions are possible and selecting from among the options which meet their needs at that scale.

This apparent connection between participation and scale is consistent with social science theory. As an illustration, consider the following line of reasoning.[3] The literature on the diffusion of innovations tells us that the most important influences on most decisions are personal communications -- not newspapers or television but face-to-face contacts (Rogers and Shoemaker, 1971; Roberts and Frohman, 1978; Sommers and Clark, 1977). Agreement is facilitated by direct human interaction; in fact, where different points of view must be reconciled or where risks are perceived to be considerable, human interaction is often a requirement.

Building on this body of research, Torsten Hagerstrand and his colleagues in Sweden have shown that there is a kind of "choreography" to human interaction (reviewed by Pred, 1977). It is shaped by the fact that both time and space limit what we can do. For example, Hagerstrand's theory of "time-geography" identifies capability constraints on interaction: There is only so much time in a day, movement is always time-consuming, etc. And it also identifies coupling constraints: if I am interacting here, I cannot also be interacting there, and a certain amount of time is required for any instance of interaction.

These kinds of constraints put bounds on what is possible in consensusbuilding at least to the extent that it depends on personal interaction. Because time is limited and each case of interaction takes time, only so much interaction can take place. And because movement takes time, the more and farther we move, the less time is available for interaction.

Such concepts lead us to the notion that, in a pluralistic social and political system, there may be limits to the social and spatial scale within which consensus or accommodation can be reached, short of an unmistakable threat. Perhaps, as Hazel Henderson has suggested (1978), a lot of our indecision about energy questions is because we are trying to deal with options whose impacts spread beyond the range that our decisionmaking structures can handle. Maybe if we want to do a better job of meeting our national needs for energy, we should focus more of our attention on the sizes of the social units that can make decisions about how to meet their own needs. Regardless how good a technology or other policy alternative may look to us energy analysts, if -- as a society -- we cannot agree to use it, then it is not helping us to solve our energy problems.

If so, we need to pay more attention both to the instinctive urge for greater self-sufficiency and to the advocacy of decentralized approaches to energy planning. From social and political points of view, energy selfsufficiency not only reduces the vulnerability of the nation; at a more detailed scale, it may offer a basis for reaching a national goal without turning back the clock on participation. Consequently, at a variety of scales it cannot be ignored as a possible energy policy objective. The question is whether it might have adverse or unsuspected economic implications.

2. ECONOMIC IMPLICATIONS OF DECENTRALIZATION AND SELF-SUFFICIENCY

A. Self-sufficiency and Decentralization

First, it is necessary to distinguish between self-sufficiency and decentralization. Clearly, there is a continuum between a condition of perfect free trade, rare between countries but not so rare between regions within a country, and autarky -- a condition of perfect self-sufficiency. Most countries fall somewhere between the poles; the typical region within a country is closer to the free trade pole than the country itself.

Self-sufficiency in energy supply is a situation where trade in energy is being limited. Some localities that in a free market would be buying their energy from elsewhere are instead producing their own. Certain localities that in a free market would be selling their energy are instead using it themselves or saving it for another day. For energy, at least, a geographical unit is located near (or moving toward) the autarky end of the continuum.

Decentralization, on the other hand, refers to where decisions are made rather than to how localities and activities are linked to each other by means of economic flows. Decentralization implies point patterns, dispersion, scatter. Self-sufficiency implies cell patterns, boundaries. The two conditions are neither corollaries nor opposites; they are different dimensions of a political economy, too closely related to be orthogonal but too independent to be highly correlated. It is possible, for example, to imagine a highly decentralized economic system in which each unit is highly specialized, highly dependent on linkages. This, in fact, is supposed to be the way the American economic system works. It is somewhat harder to conceive of a system which is centralized but consists of highly self-sufficient parts. For instance, a feudal society appears to be centralized, but in reality it is not. Decentralization, therefore, appears to be a necessary (but not a sufficient) condition for self-sufficiency; but self-sufficiency is not a prerequisite for decentralization.

B. Economic Implications of Decentralization

Since decentralization is a prerequisite for self-sufficiency, the economic implications of dencentralization also apply to a condition of energy autarky. Such implications can be drawn from studies of how to make centralized economies more efficient and from the literature on location theory, along with other scattered references.

B.1. Decentralization and Efficiency

In the 1930's, writing about the allocation of resources in a socialist economy, Hayek suggested that a principal characteristic of economic information is that it is dispersed. He further argued that the full transfer of this dispersed information to a central authority is impossible; therefore a socialist economy must be decentralized, at least to some degree, in order to operate efficiently (Hayek, 1935). This has led to sporadic flurries of discussion; for example, Marschak concluded on the basis of theory that a decentralized planning process cannot be shown to be better than a centralized process, or vice versa (Marschak, 1959). But management scientists and organization theorists have continued to be interested in the possibility that organizations are more efficient if they permit more decisions to be made by subunits. Recent theoretical work seems to indicate that decentralization is an efficient way to adapt to an environment which varies unpredictably, especially if the decision rules that apply to different parts of the organization are different (Radner, 1975).

Another possibility is that decentralization, by placing decision making closer to those who must implement decisions, is a force for actuation. For instance, Charles Lindblom has suggested that the Chinese approach to economic development is unusual in its emphasis on mobilizing the energy and resourcefulness of the masses. As he interprets their philosophy and experience, these qualities -- associated with localized coping -- can more than compensate for the benefits of organizational complexity and specialization. Motivation is substitutable for design (Lindblom, 1975).

B.2. Location Theory

One of the most elementary principles of location theory is that, if consumers are dispersed, there are good reasons for dispersing the economic activities which serve those consumers. As Figure 1 illustrates, if transport costs are high for supplying the good or service, the activity (say, energy production) will be relatively dispersed in its spatial pattern, other things equal. On the other hand, if the activity is characterized by economies of scale and agglomeration, it will tend to be concentrated in its spatial pattern. Between these two forces is an equilibrium which determines the optimal subdivision and spacing of the activities.

More precisely, it can be shown that the distribution costs of a good or service per unit of production are related inversely to the number of distribution centers. The Appendix I, for example, proves

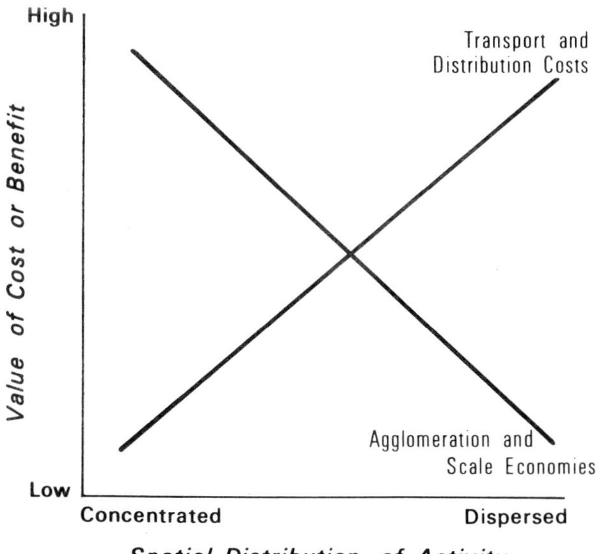

Figure 1. Spatial Distribution of Activity

theoretically that these costs decline with the square root of the number of centers; the more centers, the less the unit cost of distribution, but the savings per additional center decline as the total number of centers rises. This relationship is depicted in Figure 2, along with a curve of unit production costs which reflects economies of scale (and diseconomies of extreme agglomeration). Combining these two curves defines a total unit cost curve, showing the optimal level of dispersion.

The limitation of this derivation is that it has nothing directly to do with where decisions are made. Dispersion need not imply decentralized planning or decisionmaking. Its value, however, is that it raises two useful questions. One is whether the existing pattern of, say, electricity supply in the United States is optimal in this sense. It is clear that the recent trend has been toward larger, more remote facilities (Figure 3); some observers believe that this has taken (or is taking) us past the optimal level of concentration. This should be a researchable question.

The second issue is whether location theory might tell us something about the optimal degree of subdivision of area for a decentralized approach to energy decisions. Might it be argued that, as a boundary condition, there should not be more separate planning units than the optimal number of individual centers or facilities?

B.3. Other Implications

As an example of other insights in various literature, a book

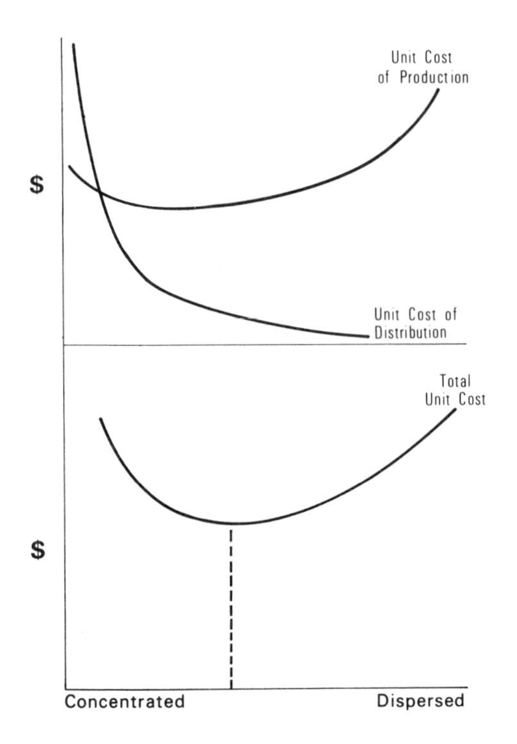

Figure 2. Optimal Dispersion of an Economic ctivity.

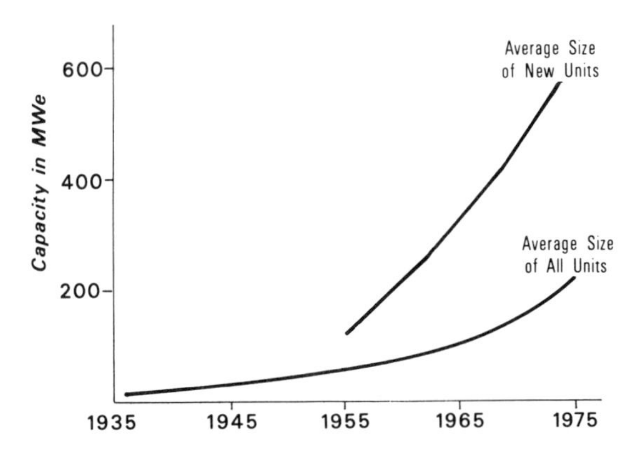

Figure 3. Trend in Average Unit Size for Electricity Generation.

applying ethics to development economies has stated that broad popular participation in decisions, which requires decentralization, must be a key principle in any development process that is truly ethical (Goulet, 1971).

These literatures on the economic implications of decentralization tell us that, at least in moderation, decentralization is good for national and regional development. It adds efficiency, it makes fuller use of available information, and it may improve motivation.

C. Economic Implications of Self-Sufficiency

In contrast, the economic implications of self-sufficiency are less attractive, although they are not all bad. The literature on trade -- especially international trade -- has explored the issues in considerable detail. Other useful insights can be drawn from studies of spatial price equilibration and relationships between economic inertia and disadvantage.

C.1. Trade Restraints

In general, economists do not like trade restraints. On the basis of both theory and observation, they believe that such restrictions lower aggregate wellbeing; overall, the costs are greater than the benefits, although the balance varies among subdivisions of the economy. Figure 4 summarizes the accepted principles, applied specifically to a tariff on imports. As for this view, "such consistent agreement is rare within the economics profession" (Kindleberger and Lindhert, 1978).

Welfare economists have gotten agreement that there are certain exceptions to this rule, where economic and social costs diverge and as a result free trade is not socially optimal, even if it is economically optimal. Given full information about such imperfections, it is possible to derive a nationally optimal tariff (a nonzero value) and to evaluate other protectionist policies as well: import quotas, export subsidies, discriminatory government procurement policies, changes in exchange rates, etc.

One of the common welfare arguments is that, when a country or a region can use trade restraints to affect the price which outsiders charge or pay for goods or services, that makes sense from its point of view. If reducing imports can drive prices down or reducing exports can drive them up, it is good for the domestic economy.

Another frequent argument is that certain kinds of comparative advantages may not develop without protection from the competition which is an aspect of free trade. For instance, perhaps a local industry that would be highly competitive once it reaches a certain level of activity or stage of development is unable to reach maturity if it must compete in its earlier stages of growth. Restraints on trade can help such "infant industries," although they appear to work better in a relatively developed economy which has resources available to support the protected activity.

A further point is that single product economies are vulnerable to

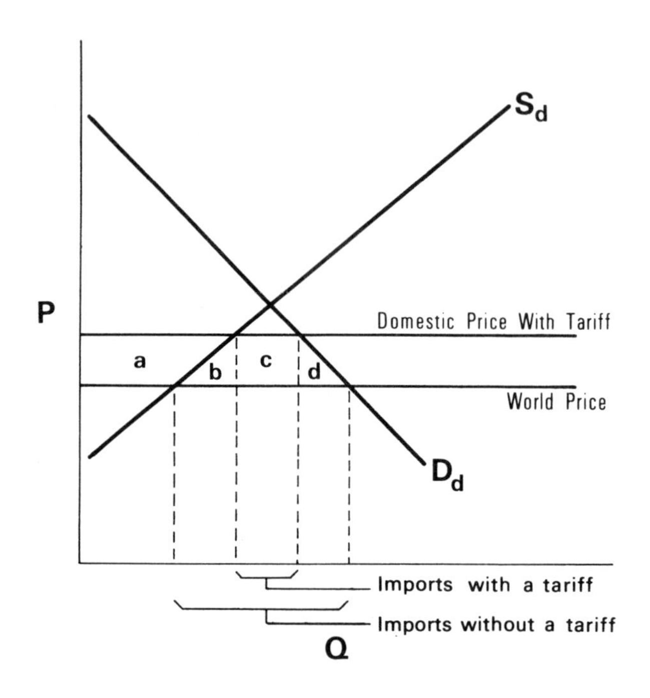

Domestic Consumers pay a + b + c + d
Domestic Producers Receive a
Government Collects c in tariff revenue

Net Loss = b - net shift by producers from cheaper imports to more expensive domestic supply
+
d - total decline in demand

Figure 4.

shifts in demand. According to portfolio choice theory, in an uncertain world one ought to diversify. In this sense, protection -- if it aids the diversification of an economy -- helps to cushion that economy against shocks from an uncertain economic environment. The net costs of protection can be considered an investment in risk aversion.

C.2. Spatial Price Equilibration

The usual theories about the costs of restraints on trade are based on partial equilibrium analyses. In other words, Figure 4 applies to restraints on the trade of energy, but it does not fully account for resulting changes in the trade of other commodities and services. More generally, however, we know from the work of Heckscher, Ohlin, Samuelson, et al., that other flows will be affected by energy self-sufficiency.Such restrictions on energy flows, whether absolute or relative, will change the energy price surface, driving up prices in certain energy-scarce regions which would otherwise be importers and lowering prices in certain energy-abundant regions which would normally be exporters. In the long run, these exaggerated price differentials will tend to disappear because other factors of production are still mobile: especially capital and labor. For example, energy-intensive activities will tend to migrate to lower energy-price regions, driving up energy prices there and reducing them in the higher price regions. This is good news for some regions, bad news for others.

C.3. Inertia and Disadvantage

An assortment of other literatures are relatively positive about the effects of restraints on trade on regional growth and development. Soviet economic planners, for instance, have argued for decades that a free market process leads to the concentration of activities at a few locations, with the result that resources elsewhere are underutilized. Because of this, they have tried to encourage complexes of activities to arise wherever the resources are. Further, after Hobson and Lenin, some observers have suggested that any system where capital is centrally controlled exploits areas outside the centers, because peripheral areas can usually get access to capital only by giving up some control over their own affairs. Finally, Myrdal and others have provided evidence that free trade favors highly developed regions, tending to perpetuate existing patterns of advantage. It distorts the trade patterns of poorer regions to benefit the rich (consider the exports of food items from LDC's to developed countries, such as fish products from Peru to Europe or maize from Thailand to Japan).

To summarize, most people believe that self-sufficiency is economically inefficient. It will mean net economic costs, disbenefits, disadvantages. On the other hand, certain restraints may be desirable for maximizing social welfare, and even when the net benefits are negative there will be winners as well as losers.

3. SELF-SUFFICIENCY AND REGIONAL ADVANTAGE

As a result, a particular region (or a particular LDC) needs to analyze how the pros and cons of self-sufficiency add up for it. On the basis of sound theory and empirical evidence, such a place will have trouble arriving at definite conclusions at this point. But while we wait for the necessary research to be done, we can start to guess what the answers will probably be, and our guesses can perhaps serve as initial hypotheses for the research effort.

To provide a foundation for the guesses, consider a future where most countries and most regions are largely self-sufficient in energy (at least self-sufficient for heating, cooling, and electricity; perhaps totally self-sufficient except for a part of the transportation sector) as contrasted with a future which is a simple extension of current trends.

A. General Pros and Cons

Within the interregional economy is a whole, we would expect the economic impacts of self-sufficiency to include both pros and cons:

1. Cons.
   a. Higher economic costs. In some regions, energy would be produced in less efficient ways than would be possible with more interregional trade; so the costs would be higher. This, in turn, would cause the economically inefficient use of other resources -- for example, inefficient substitutions of labor or capital for expensive energy. The magnitude of the extra costs would depend on many things, but they would have a positive relationship with (1) the degree of regional self-sufficiency reached and (2) the smallness of the region. As a very extreme example, a recent British study (reported by Williams, 1979) estimated that an "autarchic" single house in the UK would have seasonal heat storage costs of 11,000 pounds ($25,000). A week's supply of hot water alone would cost 300 pounds ($700). The costs in some areas could therefore be very high if the areas are very small and they try to be totally self-sufficient. A major part of the expense is assuring a perfectly reliable supply -- by means of backup systems, storage, etc. In addition to the direct economic costs of self-sufficiency, there would also be some administrative costs and some displacement costs as factors of production are shifted around.

   b. Greater interregional variations in prices. At least for awhile, we would expect an increase in variations between regions in energy prices, because energy-scarce importing regions would be substituting more expensive local energy supply or conservation, while energy-abundant exporting regions would see internal energy prices drop because of a reduction in export markets. One effect of this might be a kind of fiscal mercantilism, where energy-abundant areas offer cheap energy and relatively low tax rates (since they can avoid having to pay for a more expensive new system) to attract desirable new economic activities. This migration impact would only operate to the extent that energy prices and taxes, etc., affect location decisions (which seems to be only a fairly moderate effect), but it would be a reality. The increase in variance in energy prices would almost certainly be at least partly an inverse function of the size of the self-sufficient units.

c. Problem cases. Some special problems would result. For example, certain regions might have very few options as they try to meet their own energy needs: their internal energy resources are very limited, or their human resources are limited, or their capital resources are limited. An energy- economic system that fails to allow for such cases and to provide a way to handle them, would be exceedingly unstable, as well as unjust.

2. Pros.

a. Incentives for identifying and using local resources. With a move toward self-sufficiency, there would be greater incentives for identifying and using local resources for energy supply and local potentials for energy conservation. The question is just what these potentials are: large or small. If we are in effect underutilizing local and regional resources -- such as the sun, biomass, low-head hydropower, and conservation possibilities -- under present arrangements, a change might spring loose a whole lot of "infant resources" so that they can mature. It also might stimulate local inventiveness, as the Chinese say decentralization has done there.

b. Fine tuning of supply/use mixes. It would also make it possible to relate energy supply/use mixes pretty closely to local realities. There are a lot of differences from place to place in what makes sense in an energy system: different resources, different needs, different preferences regarding tradeoffs, different degrees of willingness or abilities to pay high prices or to accept risks or impacts. For example, Figure 5 illustrates the fact that energy prices vary considerably from county to county within the United States. An energy option to which the market is indifferent at a national average price for energy is in fact very attractive in some localities but clearly unattractive to others.

c. Security. At least initially, it would help to insure people in each region against disruptive actions or decisions by people outside; and insurance can be an economically rational investment, a benefit worth paying a price to get.

B. Speculations about Winners and Losers

The discussion above of price effects of energy self-sufficiency indicated that the impacts would differ according to whether a region is now an energy importing or exporting region. With this idea as the starting point, onc can speculate about the connection between self-sufficiency and regional advantage both interregionally and intra-regionally:

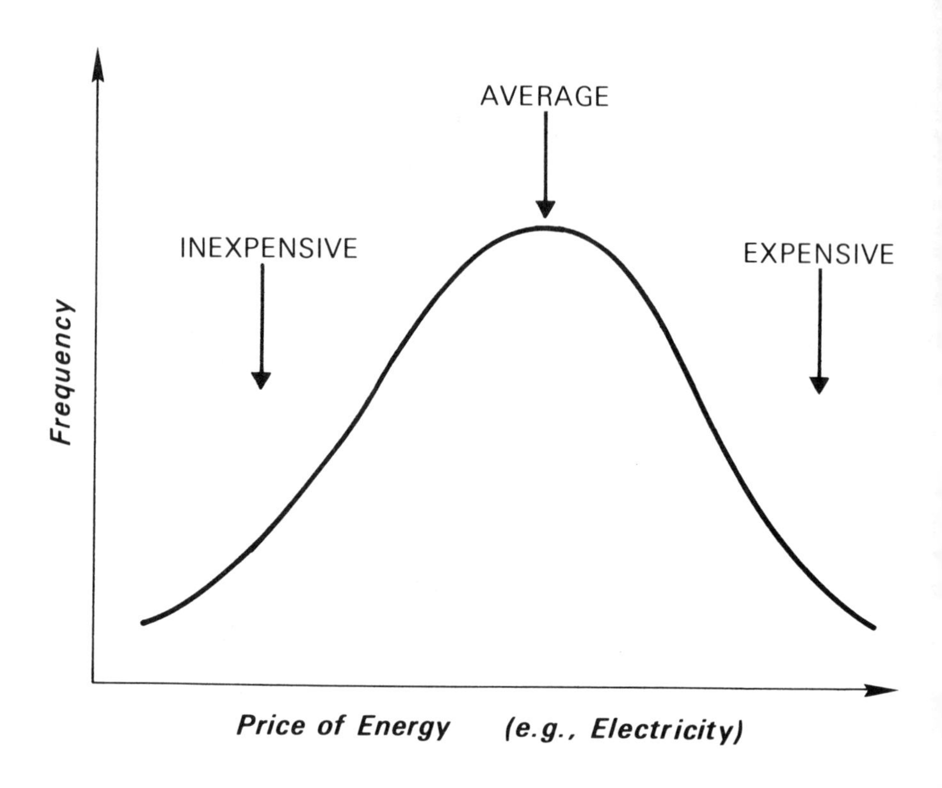

Figure 5. Energy Prices Vary From Place to Place

1.Equity among regions.

a. The effects depend on the amount of underutilized local energy resources. If the amount is small, importing regions will have few substitutes and thus high costs. Exporting regions will benefit in the long run, because they can offer abundant cheap energy to mobile industry. If, however, the amount is large, it is likely that many importer needs can be met with substitutes at a reasonable price, which hurts the exporting regions by reducing the demand for a part of their basic economy.

b. The effects depend on the spatial distribution of the underutilized local energy resources. If they are widely dispersed, the effect will probably be to increase interregional inequalities, because the wealthier regions will be better able to take advantage of their local resources. If the underutilized resources are concentrated in location -- but located in a way unrelated to current patterns of wealth -- the effect would probably be to reduce inequalities, because some non-rich regions would likely benefit while some rich regions would not.

c. The effects depend on the enrgy intensity of the regional economy. The economic effects of self-sufficiency would be less intense in regions with lower energy intensity because, for instance, energy price effects would be less severe for their macroeconomies. This might be an advantage for less wealthy regions.

2. Equity within regions.

a. Higher energy prices would tend to increase inequalities. In the United States, energy expenditures are a larger part of the family budget for poor people than rich; thus higher prices add to income inequities, unless special remedial policies are put into place. The same is probably true for most other countries as well, except for certain LDC's where the poorest groups meet most of their energy needs by non-market gathering of sun, wood, and wastes.

b. The effects depend on institutional structures. The ownership and organization of energy supply and use facilities would almost certainly be different in a self-sufficient region, and this shapes income distribution and other aspects of equity. Freedom from external domination is very attractive, but it is worth remembering that through history local control has usually strengthened the hand of local power structures. It has not necessarily led to a more equitable economic and political system within regions.

c. The effects depend on the change in energy import-export levels. Reversing The Stolper-Samuelson Theorem (which concerns the effects of a change from self-sufficiency to free trade), one would expect that with self-sufficiency the returns to factors of production involved in energy sector activities (including labor) would rise in formerly importing regions and fall in formerly exporting regions. The question is whether improvements in wages in the energy sector will lessen or increase inequalities within the region.

d. In ways that are self-evident, the effects depend on the pattern of ownership and employment in energy-related activities.

Overall, it appears that a high degree of national and regional self-sufficiency in energy supply and use would probably be expensive, problematic, and disruptive to the growth plans of many regions unless either (a) considerable amounts of underutilized local resources are available in most places (one should not underestimate the chances of this) or (b) a variety of policies are introduced to counteract the adverse impacts of self-sufficiency.[4]

In terms of world energy problems, of course, there are other issues as well, related to the sheer magnitude of the economic constraints on energy policies in the LDC's. Particular concerns must be with (1) the possibly catastrophic effects of a failure to increase relatively cheap energy supplies by whatever means (does a move toward self-sufficiency mean smaller quantities of affordable energy ahead, even if only for a few decades?) and (2) the capital and other resources required to develop a self-sufficient energy system. The real challenge, and it is one of the greatest of our times, is to respond to the concerns that lie behind the attractiveness of self-sufficiency and, at the same time, to bring about a massive increase in energy supply to the LDC's at a price that facilitates economic and social development.

4. CONCLUSION

The paper began with the question: is energy self-sufficiency good for regional and national development or is it bad? Without an abundance of research that is yet to be done, the answer remains uncertain; but several general truths seem clear:

(1) Energy self-sufficiency would be good for some places but bad for others.

(2) Relative self-sufficiency is unlikely to be bad for the development of areas with plentiful energy resources -- either presently known or presently underestimated because energy systems operating at a more general scale have paid them too little attention -- as long as those areas can get the capital necessary to develop and use their resources.

(3) Complete self-sufficiency is illusory, because energy sufficiency will always be related to interdependence with regard to other factors of production.

(4) Energy self-determination makes sense even when self-sufficiency does not.

(5) But even self-determination can mean social costs to our international society and economy if we fail to provide mechanisms to deal with special problems, such as constraints on capital flow, the needs of particular areas with a paucity of resources, or requirements for conflict resolution.

Because shifts toward greater self-sufficiency can be expected during the next several decades, at least at the national level, it is rather urgent that we move beyond these kinds of truisms to a more careful, well-documented understanding of what this might mean for regions, countries, and our international political economy as a whole. Otherwise, we run the danger that a solution to one kind of problem will exacerbate others that are more fundamental and lasting.

NOTES:

[1] The conclusion was that indeed it could, except for some liquid fuel for transportation

[2] Not to mention passing such "partnership" legislation as ECPA, EPCA, EMPA and EECA.

[3] Others include Mumford and others on city size, the political science literature on federalism, Fisher and Etzioni on "fractionation" as an approach to conflict resolution, the work of ecologists and anthropologists (notably Roy Rappaport) on characteristics of stable systems, and the work of economists and management scientists on organizational efficiency (e.g., Radner, 1975).

[4] For instance, if the focus is on regional and local self-determination rather than self-sufficiency, many of the advantages are retained without many of the disadvantages (Wilbanks, 1980).

APPENDIX

DISPERSION AND DELIVERY COSTS*

(1) Define ϕM_j as total delivery costs with $_j$ facilities.

(2) Assume that the cost of delivery for any facility F_i is proportional to the average distance from the facility to a point in its service area R_i which has dimensions ℓ_i and b_i.

(3) If F_i is centrally located in R_i

Then:

$$\iint_{R_i} \sqrt{x^2 + y^2} \quad dxdy \quad = \int_{-\ell_i}^{\ell_i} \int_{-b_i}^{b_i} \sqrt{x^2 + y^2} \quad dxdy$$

If R_i is replaced by four areas of equal size bisecting each side, with one facility replaced by four facilities, then:

$$= 4 \int_{-\frac{1}{2}\ell_i}^{\frac{1}{2}\ell_i} \int_{-\frac{1}{2}b_i}^{\frac{1}{2}b_i} \sqrt{x^2 + y^2} \quad dxdy$$

Letting $2x = z$, $2y = w$, we obtain:

$$= 4 \int_{-\ell_i}^{\ell_i} \int_{-b_i}^{b_i} \tfrac{1}{2} \sqrt{z^2 + w^2} \ \tfrac{1}{4} \ dzdw$$

$$= \tfrac{1}{2} \int_{-\ell_i}^{\ell_i} \int_{-b_i}^{b_i} \sqrt{z^2 + w^2} \quad dzdw$$

So: $\phi M_4 = \tfrac{1}{2} \phi M_1$

Similarly:** $\phi M_k = \frac{1}{\sqrt{k}} \phi M_1$

---

*This approach was suggested by A. Schwartzkopf, Department of Mathematics, University of Oklahoma.

**At least where $\sqrt{k}$ is an integer.

BIBLIOGRAPHY

Branscomb, L., Testimony before a joint hearing of the U.S. Senate Committee on Commerce, Science, and Transportation and the U.S. House Committee on Science and Technology, February 1978.

Cecelski, E., J. Dunkerley, and W. Ramsay, 'Household Energy and the Poor in the Third World.' RFF Research Paper R-15, Resources for the Future, Washington, D.C. (1979).

Goulet, D., The Cruel Choice, New York: Atheneum. (1971).

Haggett, P., Geography: A Modern Synthesis, 2nd ed., New York: Harper and Row. (1975).

Hayek, F. von, ed., Collectivist Economic Planning, London: George Routledge and Sons. (1935).

Henderson, H., 'Science and Technology: The Revolution from Hardware to Software,' Technological Forecasting and Social Change, V. 12, pp. 317-324. (1978).

Islam, M.N., Informal comments at an International Workshop on Energy Survey Methodologies for Developing Countries, sponsored by the Board on Science and Technology for International Development, NAS-NRC, January. (1980).

Kash, D.E., and others, Our Energy Future; The Role of Research, Development, and Demonstration in Reaching a National Consensus on Energy Supply, Norman: University of Oklahoma Press. (1976).

Kindleberger, C. P. and P.H. Lindhert, International Economics, 6th ed., Homewood, Illinois: R.D. Irwin (especially pp. 107-147). (1978).

Lawrence Berkeley Laboratory, Distributed Energy Systems in California's Future, Interim Report, 2 v. (Washington: U.S. Department of Energy, Office of Technology Impacts, March). (1978).

Lindblom, C. E., 'The Sociology of Planning: Thought and Social Interaction,' in M. Bronstein, ed., Economic Planning, East and West, Cambridge, Massachusetts: Ballinger. (1975).

Marschak, T., (1959). 'Centralization and Decentralization in Economic Organizations,' Econometrica, V. 27: pp. 399-430. (1959).

Mbi, E., 'Energy Surveys in Urban Areas: Experiences in Burundi and Cameroon,' excerpted in Proceedings, International Workshop on Energy Survey Methodologies for Developing Countries, NAS-NRC, Washington, D.C.: National Academy Press. (1980a).

Mbi, E., Informal comments at workshop, Ibid. (1980b).

Murray, F. Y., ed., The National Coal Policy Project: Where We Agree, Center for Strategic and International Studies, Georgetown University, 2 v., Boulder, CO: Westview Press. (1978).

Pred, A., 'The Choreography of Existence,' Economic Geography, v. 53: pp. 207-221. (1977).

Radner, R., 'Economic Planning Under Uncertainty,' in M. Bronstein, ed., Economic Planning, East and West, Cambridge, Massachusetts: Ballinger. (1975).

Roberts, E. C., and A. L. Frohman, 'Strategies for Improving Research Utilization,' Technology Review, v. 80, no. 5 (March/April): pp. 33-39. (1978).

Rogers, E.M. and F.F. Shoemaker, Communication of Innovations, New York: Free Press. (1979).

Schurr, S.H. et al., Energy in America's Future, Baltimore: Johns Hopkins. (1979).

Sommers, P. and R. W. Clark, 'ERDA and the Diffusion of Innovations,' Yale Mapping Project on Energy and the Social Sciences, New Haven: Institute for Social and Policy Studies, Yale University. (1977).

State of Oregon, 'Oregon's Energy Perspective,' PB-224146, State Executive Department, Salem, Oregon, May. (1973).

U.S. Department of Energy, Joint Egypt/United States Report on Egypt/United States Cooperative Energy Assessment, 5 v., Washington, D.C.: U.S. Government Printing Office. (1979a).

U.S. Department of Energy, Joint Peru/United States Report on Peru/United States Cooperative Energy Assessment, 4 v., Washington, D.C.: U.S. Government Printing Office. (1979b).

U.S. Department of Energy, Reducing U.S. Oil Vulnerability, DOE/PE0021, Washington, D.C.: U.S. Government Printing Office. (1980).

Wilbanks, T.J., 'Decentralized Energy Planning and Consensus in Energy Policy,' paper presented at the National Energy Policy Conference, Morgantown, West Virginia, May. (1980).

Williams, K., 'Future Role and Availability of Hydrocarbons,' presentation at the Oak Ridge National Laboratory, January 31, 1980. (1979).

II Systems and Models for Energy-Environmental Assessment in the Netherlands

5 An integrated Environmental Model for Regional Policy Analysis

J. W. ARNTZEN AND L. C. BRAAT

1. INTRODUCTION

Environmental problems have most often been studied in a monodisciplinary fashion. Many facts and insights regarding various aspects of these problems have been obtained in that way. However, the relationships between aspects of the problems, which are fundamental to the solutions, are often neglected. In order to provide policy-makers with comprehensive environmental analyses, integration of monodisciplinary studies is imperative.

The usefulness of models in analyzing complex problems has been widely recognized. By nature, multidisciplinary environmental analysis is complex, hence requiring the application of models. In this paper the multidisciplinary approach is exemplified in an integrated environmental model.

In the past ten years integrated environmental studies have been initiated by both scientific institutions, environmental action groups and several departments and agencies in administrations of various countries, (Heinzmann, 1980; Rijn, 1976/77; and Thoss, 1978).

At the Free University of Amsterdam, a cooperation between the Institute for Environmental Studies and the Department of Economics has resulted in two theoretical papers regarding environmental modeling (Nijkamp and Opschoor, 1977; Arntzen, et al, 1978).and subsequently in a research project for which funding has been provided by the National Physical Planning Agency, (Arntzen et al., 1978, 1981).

This project, which started in January 1979, aims to develop a regional integrated environmental model which should function as a device for assessing economic, social and ecological effects of regional policy. To build a general model, which can be used for various types of policy-analysis and in different regions, first, a specific model is built. In the present set up , demographic, economic and ecological aspects and urban and regional facilities are integrated within a spatial framework. Several aspects of regional environmental problems have been excluded from further analysis in this approach. Health and social wellbeing of human population are not yet included in the model, due to lack of research capacity. An integrated environmental model may be used to assess effects of regional policy in general, but also more specifically to put Environmental Impact Statements in a regional perspective and to analyze long term feedback effects of particular economic and social activities.

2. STRUCTURE OF THE MODEL

In environmental problems, both quantitative and qualitative aspects are of importance. The latter ones are often not encompassed in models. In the model presented here, qualitative variables are explicitly included.

In the present state of development, the Integrated Environmental Model (IEM) consists of five submodels. A demographic submodel generates size and composition of the population of the region. The demand for, the use and the capacity of facilities for social and economic activities are brought together in a facilities submodel. The economic cycle between demand and supply appears to be hardly quantifiable at a regional scale because of lack of data about consumption and regional leakages (import to and export from the region). The economic submodel therefore concentrates on the supply side, the production of goods and services.

The components and processes of the natural environment are described in the ecological submodel. This submodel contains several sets of variables, each set representing a different ecological system (e.g., marine, fresh water and various terrestrial ecosystems). Some relations between the natural environment and human society are described and analyzed in this submodel by the concept of functions of the natural environment. This concept is defined as "the supply of means for human use", (Braat, et al., 1979). The word supply includes both the actual and the potential supply of means.

The performance levels of the functions offer a measure of the relations between ecological systems and economic and social systems.

The fifth submodel does not consist of a set of related variables, but of an aggregate of procedures, formulae and data sets (e.g., maps). In this submodel three aspects of space are described: surface area, location and distance. This intermediate submodel functions as a device to integrate spatial aspects into the interactions between variables of the other submodels. In Figure 1 the submodels and the major relationships within the Integrated Environmental Model are shown.

FIGURE 1. STRUCTURE OF THE MODEL

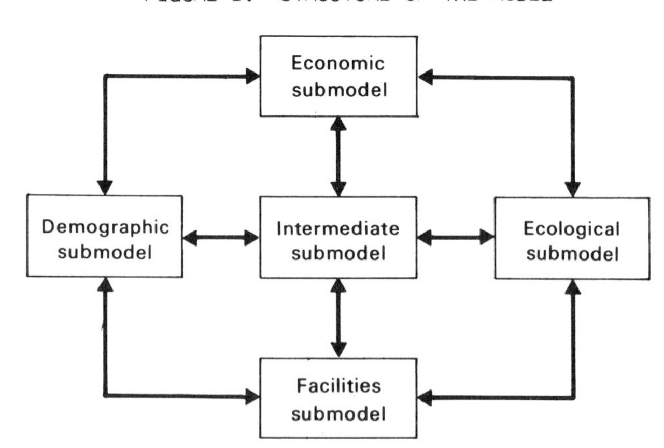

To construct a generally applicable model, two steps have been distinguished. First a model is developed, based on a particular plan for a specific region in the Netherlands involving changes in population, economy, facilities, land use patterns and ecological systems.

Secondly, the specific model that results is generalized in order to make it applicable to other regional plans and other regions. Therefore, the potential for generalization of data banks, variables, equations and methodology is considered in making choices for the development of the specific model.

Integration is the basic characteristic of the study. Variables for the submodels have been selected according to the following criteria: relation to the plan, relevance for regional planning, availability of data and function in the structure of the (sub)model. Subsequently, equations are specified, which describe the relations between the variables.

Not all relationships can be fitted in mathematical equations. Those cases, where dose-effect relationships are not quantitatively known and where subjective factors are important, like in the perception of the attractivity of an area to live, to work or to recreate in, are described in qualitative terms.

A particular feature of this study is that it aims to integrate existing data, rather than to rely on extensive field work. Many problems are met in attempting to integrate data from different sources and collected for different objectives. Because the model aims to be a device for policy-analysis, accessibility of the model for the policy-maker and for the public has high priority. This and the subjective and qualitative aspects embodied in the model suggest to develop an interactive model. Decisions about details of policy, priorities and boundary conditions are therefore chosen to be part of the analysis.

The model is being developed at the regional scale, which in the Netherlands averages about 1000 km^2. This scale has been selected, because it is widely used by planners and policy-makers and because it offers a possibility to integrate economic, demographic and ecological data. This compromise scale requires disaggregation of economic data (usually available at the national or provincial level), aggregation of demographic (available at the local level) and ecological data (available at local and sub-local level).

The time period for which the effects of a regional plan are analyzed in the Integrated Environmental Model is another compromise. The period chosen (10-15 years) is a compromise between accuracy of the output of the model and relevance for the planners and policy makers.

3. THE PLAN

The specific model is being developed for a region in the Southwestern part of the Netherlands in which an urban development plan is designed. This plan is an example of the national policy to protect open space in the landscape and concentrate urban development in selected urban areas. The main element of the plan is the construction of 10,000 houses within a period of 10 years in a polder, located in the Eastern part of an estuary adjacent to the city of Bergen op Zoom (see Figure 2).

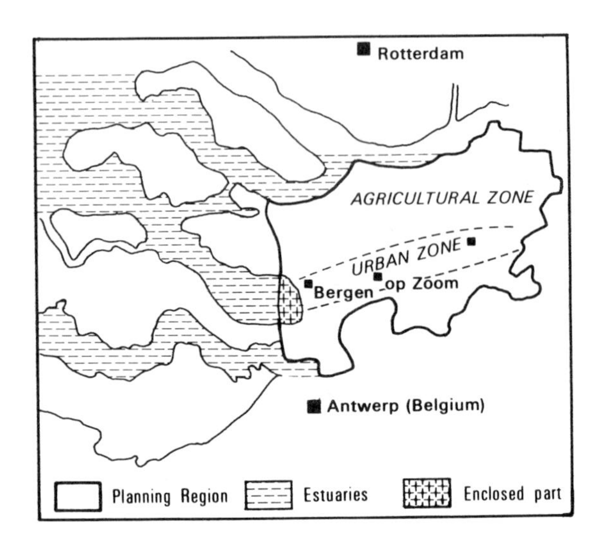

Figure 2. The region West-North Brabant

Part of the estuary is enclosed by dikes and here approximately 300 ha. is to be drained to become a polder. Directly related to the plan are an increase of industrial production capacity, allocation of part of the houses to immigrants (from the Rotterdam region), construction of recreational facilities and increase of energy use and other utilities.

4. THE REGION

The region West-Brabant is quite diverse, both as to the artificial and the natural environment. Agricultural land use dominates the region (71.7%). Natural areas cover about 12,000 ha. (approximately 10%), while urban areas (including industrial plants and roads) occupy approximately 18% of the total surface. The population has been increasing at 16.8^{0}/oo (in 1976), but this is mostly due to immigration from the Rotterdam region. The unemployment rate is somewhat higher than the national average.

The water quality in the region varies from good in the smaller creeks and rivers to very bad close to the industrialized areas. The air quality is strongly influenced by the extensive industry in Antwerp (Belgium) and Rotterdam.

5. THE SUBMODELS

The construction of the model has started with the selection of the variables for each of the submodels. The criteria for selection have been mentioned above. In the demographic submodel, population is classified by age and sex. This submodel operating at the municipal scale, generates forecasts of the size and composition of the populations for any aggregate of municipalities. These forecasts are used for the economic submodel (labor market) and the facilities submodel. The latter submodel contains those facilities which are of particular interest to the natural environment of the region. This study is not concerned with aspects of the natural environment in urban areas (e.g., parks).

Therefore, urban facilities are aggregated into one category. Other categories include facilities for water resources, waste disposal, recreation and energy. Several aspects of the facilities are distinguished in this submodel: 1) household demand and use, 2) total demand and use (including demand and use of industry, generated in the economic submodel) and 3) supply capacity and related surface area. In the economic submodel the bottom up approach is used. Production functions are specified for nine production sectors. However, for adequate analysis of pollution, resource use (water and energy) and production of waste, a more detailed classification is needed. Whenever necessary, a standard classification up to 300 sectors can be applied. The development of employment in the region is described by a policy scenario (for the 'plan situation') and a trend-scenario (for the 'reference situation'). In the ecological submodel, the enclosed estuarine system has been modeled first, because this system is most directly affected by the plan. Aspects of water quality, vegetation and fauna are used as indicators of changes in the system. Feedback relations exist via the assessment of the condition of the system by recreationists and indirectly with the local economy (expenditures by recreationists). Feedback relations exist also through a set of standards for air and water quality and nature conservation objectives. These relations require that the qualitative aspects of the system are described, too. In the intermediate submodel, surface areas of land use systems have been registered and a land use map is designed based on these data.

The surface areas of variables in the other submodels are aggregated in the intermediate submodel per land use system. Changes in surface area and location, which are forecasted by the model, are either allocated by the policy-maker or by the analyst according to an explicit set of standards and rules of dominance for land use systems.

Another aspect of space-distance-is important in determining the concentrations of pollutants in air and water. In general, concentrations decrease with distance, which implies that estimates of air and water quality and effects on organisms must take account of the distance factor. Various models describing and predicting diffusion of pollutants are tested for inclusion in the integrated environmental model, (Janssen, 1979).

6. VARIABLES AND EQUATIONS

The variables that have been selected are listed in Table 1. Two major groups are distinguished: endogenous and exogenous variables.

The first category contains those variables that are explained within the model, while variables of the second group ar not influenced by other variables of the model.

The integrative character of the model is mainly determined by the variables in each submodel which are related with variables of other submodels. Integration of the monodisciplinary submodels is actually effectuated by those equations and descriptive relationships, in which variables of different submodels are related to each other. In fact, the interactions between the economic, social, demographic, ecological and facilities structure of the region form one of the criteria for selecting variables. The submodels also contain variables which are only related with variables in the same submodel. These variables are selected either for internal consistency of the model, or for descriptive or for indicator functions.

The availability and quality of the data determine to a large extent whether a relation can be translated into a mathematical equation or has to be described in qualitative terms. The endogenous variables in Table 1 are listed with their respective explanatory variables. The relationships are presented in a descriptive way and characterized as to type of equation by which they may be described.

Since the model is being designed for policy-analysis, special attention is paid to variables, which describe components, processes and activities which can be controlled by policy-makers. In the present study, this attention is focused on those aspects of the region which are within the realm of regional planning agencies.

The information presented in Table 1 is illustrated in the next section in the analysis of effects of the plan.

7. ANALYSIS OF THE EFFECTS OF THE PLAN

The major elements of the plan have been mentioned above. The effects of the plan are assessed in comparison with a reference situation for the region. In this reference situation the polder is not constructed, the extra 10,000 houses not built, etc.

The analysis starts with the introduction of the change in the initial values of the variables involved. Next, simulations with the model equations are carried out. In Figure 3 a diagram of the affected variables is presented, which is based on Table 1. The starting points of the various elements of the plan in the model are indicated. The following effects can be traced:

a. Effects of the polder

The construction of the polder leads to changes in land use (intermediate submodel) and in the surface area of the aquatic and terrestrial systems of the enclosed part of the estuary. Consequently, water volume, salinity and phosphate concentrations are affected, which in turn influence biomass and species composition of the ecosystems. Next to these changes in ecological variables, changes in variables of other submodels are caused by this land use alteration. The polder and the adjacent urban lake (between the polder and the old city) are

TABLE 1
VARIABLES AND RELATIONS OF THE MODEL

Exogenous Variables

Net migration
Net commuters
Production of houses
Surface area (s.a.) of houses
Water supply capacity
S.a. water supply capacity
Waste processing capacity
S.a. waste processing capacity
Electricity production capacity
Electricity distribution capacity
S.a. electricity capacity
Recreational facilities
S.a. recreational facilities
Sewage treatment capacity
S.a. sewage treatment capacity
Investment
Employment
Regional leakages
S.a. aquatic ecosystems
S.a. terrestrial ecosystems
Water volume
Phosphate-input and -output
Salinity-input and -output
Ground water input
Migration of animal species

Endogenous Variables

Dependent Variables	Explanatory Variables	Characteristics		
Population size	Population size, migration	A	D	F
Labor force	Population size	A	D	F
Labor supply	Labor force,commuters	A	D	G
Demand for houses	Population size	B	D	F
Stock of houses	Stock of houses, production of houses	A	D	G
S.a. houses	Stock of houses	A	D	F
Water use, households	Population size	B	D	F
Water use, total	Water use households and industries	B	D	G
Supply ground water	Ground water stock,input	B	D	G
Waste households	Population size	B	D	F
Waste, total	Waste households and industries	B	D	G
Electricity use households	Population size	B	D	F
Electricity use, total	Electricity use households and industries	B	D	G
Effluent households	Population size	B	D	F
Air pollution household	Population size	B	D	F

Dependent Variables	Explanatory Variables	Characteristics		
Recreationists	Population size	B	D	F
Recreation density	Recreationists, recreation facilities and areas	B	D	F
Traffic	Population size, production volume	B	C	
Production volume	Labor supply,investment	B	D	E
Value added	Production volume	A	D	F
S.a. production	Employment	A	D	F
Air pollution industries	Production/Employment	A	D	F
Effluent industries	Production/Employment	A	D	F
Water use industries	Production/Employment	A	D	F
Electricity use industries	Production/Employment	A	D	F
Unemployment	Labor supply, employment	B	D	F
Consumption	Population size, value added	B/A	D?	E
Phosphate-concentration	Input Phosphate, water volume	A	D	F
Salinity	Input S, water volume output S	A	D	F
Biomass aquatic vegetation	Biomas aquatic vegetation, concentration P, s.a. aquatic ecosystem, recreation density	B	D	E
Biomass terrestrial vegetation	Biomass terrestrial vegetation, ground water level, s.a. terrestrial ecosystem, recreational density	B	D	E
Biomass aquatic fauna	Biomass aquatic fauna, water volume, biomass aquatic vegetation, recreation density	B	D	E
Biomass birds	Biomass birds, biomass aquatic fauna and vegetation, recreation density	B	D	E
Species composition aquatic vegetation	Salinity, concentration P, recreational density, migration of species	B	C	
Species composition terrestrial vegetation	Ground water level recreational density, migration species	B	C	
Species composition aquatic fauna	Salinity, species composition, aquatic vegetation, migration species	B	C	
Species composition birds	Species composition aquatic and terrestrial vegetation, species composition aquatic fauna	B	C	

Dependent Variables	Explanatory Variables	Characteristics		
Ground water stock	Ground water stock, input ground water, output	B	D	E
Urban land use	S.a. houses, s.a. recreational facilities	B	D	G
Infrastructural land use	Traffic	B	D	G
Industrial land use	S.a. production	B	D	G
Public facilities land use	S.a.:sanitary purposes, electricity capacity, sewage treatment, incineration	B	D	G
Resource extracting land use	S.a. ground water extraction	B	D	G
Agricultural land use	Agricultural production	B	D	G
Land use aquatic ecosystems	S.a. aquatic ecosystems	B	D	G
Land use terrestrial ecosystems	S.a. terrestrial ecosystems	B	D	G
Total land use	Urban land use, infrastructural, industrial, public facilities, resource extraction, agricultural, aquatic ecosystems, terrestrial ecosystems	A	D	G

A = relations between variables of one submodel
B = relations between variables of different submodels
C = qualitative relations
D = quantitative relations
E = behavioral equations ($Y=f(x_1,..x_n,a_1,..a_n)$; $a_1,..a_n$ has to be estimated)
F = definition ($Y=f(x_1,..x_n,a_1,..a_n)$;$a_1,..a_n = 1$)
G = identity ($Y=f(x_1,..x_n,a_1,..a_n)$;$a_1,..a_n = 1$)
s.a = surface area

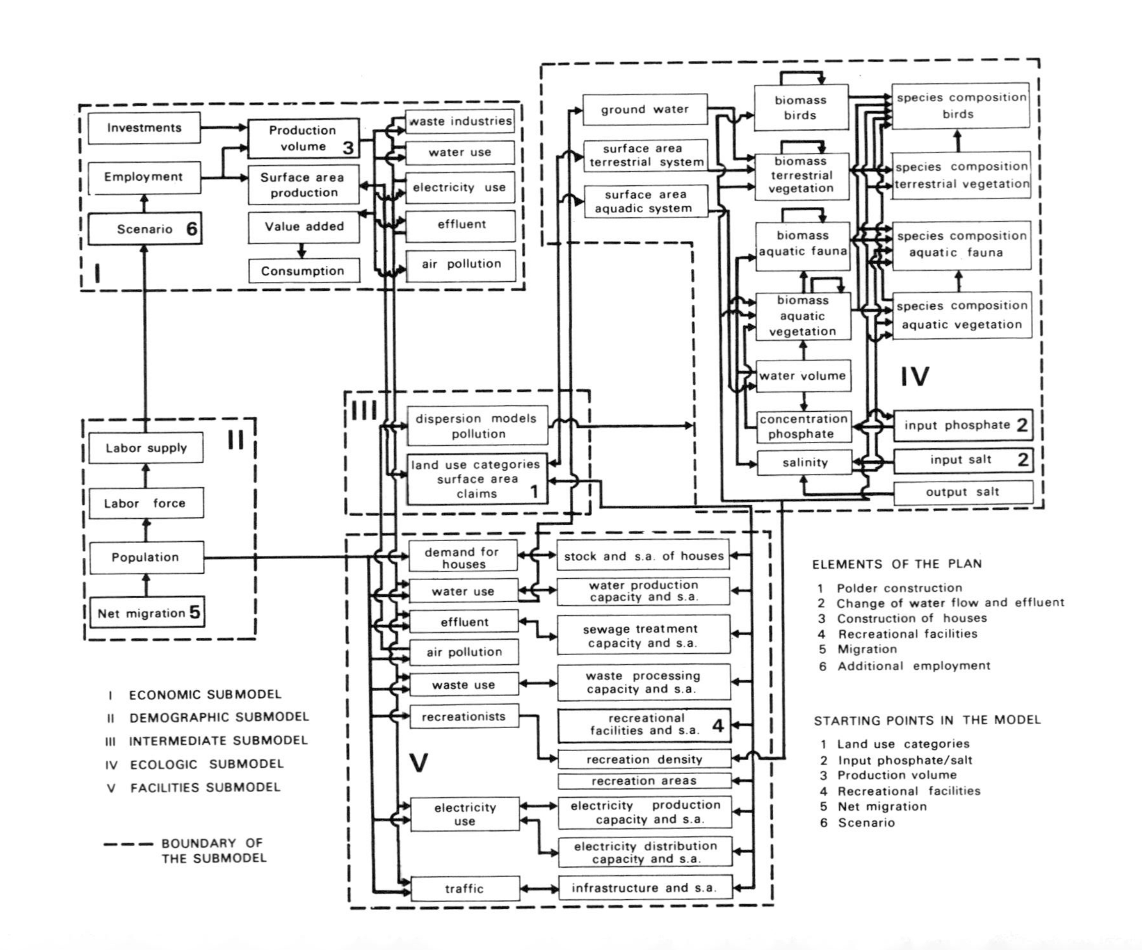

Figure 3. Effects of the Plan.

classified as urban land use, which causes a change in that category. The enclosed part of the estuary then consists of a polder, an urban and a "natural" lake. The latter one may function as a recreational area, with a potential increase of disturbance of developing aquatic and terrestrial vegetation and animal species and pollution of the water. Altogether, the amenity of the landscape and constituting systems is altered.

b. Effects of changes in water flow and inputs of effluent

The construction of the polder allows for a transfer of polder and city effluents from the natural lake to the urban lake. Also, the water level of the natural lake can be regulated with an input of urban lake water instead of water from the adjacent part of the estuary, which is more polluted and has a higher salinity.

Consequently, the natural lake will be desalinated faster than without the creation of the polder and the urban lake. Moreover, the lake will be less polluted. This may result in a greater attractivity of the natural lake for wildlife and recreational purposes.

c. Effects of the construction of houses in the polder

The construction of houses results in extra activity in the building and construction sector (economic production) and in an increase of the stock of houses (facilities submodel). No direct change in land use is registered since the polder has already been designated as urban land use. Additional urban facilities can be located inside the city of Bergen op Zoom. The surface area necessary for these facilities is included in the surface area of houses.

d. Effects of recreational facilities

Construction of the facilities leads to effects comparable to the contruction of the houses, but additionally a change in land use is caused by this element of the plan (terrestrial ecosystems are transformed with urban land use). The facilities may stimulate recreational activities, and alter the pattern of recreation and the recreation density in the region. Ecological effects have been described above (see 1, effects of the polder); a local economic effect is the increase of expenditures by recreationists and the increases of employment.

e. Effects of migration

According to the plan approximately 20,000 immigrants will enter Bergen op Zoom. A direct effect is the increase in the population of the region. Indirect effects are: an extra demand for houses (largely satisfied by the increase in supply of houses (Arntzen et al., 1978,1981)), increase of air and water pollution by households, extra demand for water (with effects on the ground water stock and ecological systems depending on it), extra garbage and sewage, increase of recreational activities and of the labor force.

f. Effects of additional employment

According to the plan, the increase of the labor force is met by an equal increase of regional employment ("policy scenario"). The latter presupposes an expansion of production capacity, production and surface area of production. When the area which is presently allocated to industrial land use, is no longer sufficient, a reallocation of space may become necessary (depending on political priorities) involving changes in other land use systems (accounted for in the intermediate submodel). Other effects of additional employment are those related to the increase in production volume (water and energy use, air and water pollution). To analyze these effects at a regional scale, a detailed classification of the production sectors has to be used. Emission of pollutants and resource extraction cause effects which via the intermediate submodel are described in the ecological submodel, while in case of water production for industrial use, water processing facilities (facilities submodel) are involved too.

8. PERSPECTIVES

In this paper a general outline of the Integrated Environmental Model has been summarized. It is obvious that the framework presented here needs to be elaborated before the model becomes fully operational in environmental policy analysis. Three different sets of activities are distinguished in completing and operationalizing the model.

First, the relations must be specified and tested. Step-by-step submodels need to be linked to each other and finally simulations with the entire model have to be carried out. However, some submodels are incomplete, as to variables, relations and connections with other submodels. For instance, a dispersion model for pollution must be incorporated in the intermediate submodel.

In addition, other eco-systems, which are indirectly affected, have to be described in the ecological submodel. Next, the specific model which is developed in the first stage of the study must be generalized. By comparing the structure of other regions with the structure of the West-Brabant region, the degree of generality of the model can be assessed and adaptations for general applicability determined.

Another aspect of generalization is the applicability of the model to different types of plans. The degree of generality of the specific model will be determined by an inventory of the components, processes and activities which are directly affected by the various plans. The third stage concerns a further elaboration and addition of aspects of the environmental problems and of the regional structure, which for various reasons have not yet been incorporated in the model. For example, environmental aspects of energy production, distribution and consumption have been included only partly in the model. However, it can be expanded within the framework. Social and health aspects may also be added to the present set of submodels.

Finally, it is the authors' feeling that the integration of different disciplines involved in environmental problems, is possible. We encountered many problems and experienced a lot of the communication problems between the different fields of research. Nevertheless, we hope that our contribution is of value for tackling environmental problems in the way it should be done, viz. in a multi-disciplinary way.

BIBLIOGRAPHY

Arntzen, J.W., P. Nijkamp, J.B. Opschoor, Geintegreerde Milieumodellen, Wenselijk en Mogelijk? ESB, 5-7-1978.

Arntzen, J.W., L.C. Braat, L. Hordijk and S.W.F. van der Ploeg, Geintegreerd Milieumodel, een Theoretisch Kader voor het Toetsen van Ruimtelijke Plannen, Institute for Environmental Studies, 79-5, Free University, Amsterdam. (1979).

Arntzen, J.W., L.C. Braat, F. Brouwer and J.P. Hettelingh, Geintegreerd Milieumodel, Final Report, Institute for Environmental Studies, Free University, Amsterdam, (1981).

Braat, L.C., S.W.F. van der ploeg, F. Bouma, Functions of the Natural Environment, an Economic-Ecological Analysis, Institute for Environmental Studies, 79-9, Free University, Amsterdam, (1979).

Heinzmann, J., 'A Complex Regional Strategy Needs a Complex Regional Analysis,' Institute of Geography and Geoecology of the Academy of Science of the German Democratic Republic, Berlin. (1980).

Janssen, L.A.M., Rekensysteem Luchtverontreiniging II, TNO-CMP 79/1, Delft, (1979).

Nijkamp, Peter and J.B. Opschoor, Naar een Geintegreerd Milieumodel? Institute for Environmental Studies, A-18, Free University, Amsterdam, (1977).

Rijn, H.T.U. van, Geintegreerd Milieuonderzoek: een Onderzoek naar de Seleving van het Ruimtelijk Milieu in het Rijnmondgebied, Openbaar Lichaam Rijnmond, Rotterdam, (1976/77).

Thoss, R., 'Regional Environmental Information Systems', Paper for the International Symposium on Regional Dimensions of Environmental Policy, Berlin, (1978).

6 An Interregional Policy Model for Energy-Environmental Management

P. LESUIS, F. MULLER AND P. NIJKAMP

1. INTERREGIONAL INPUT-OUTPUT MODELS

In a rather simple way, a regional input-output model may be represented (Lesuis et al., 1980; Nijkamp, 1979) as:

$$(I - T^r)\,\underline{x}^r = y^r \quad , \qquad (1)$$

with $\underline{x}^r$ a vector of gross production in region r[1]), T^r a matrix of regional input-output coefficients of region r, $\underline{y}^r$ a vector of final demand for regional products from region r, and I a unit matrix.

In complete interregional input-output model the trade relationships between the regions should be explicitly specified. For two regions, k and l, the following interregional input-output model may be proposed:

$$\begin{bmatrix} I-T^k & -M_l^k \\ -M_k^l & I-T^l \end{bmatrix} \begin{bmatrix} \underline{x}^k \\ \underline{x}^l \end{bmatrix} = \begin{bmatrix} \underline{y}^k \\ \underline{y}^l \end{bmatrix} , \qquad (2)$$

here M_l^k denotes the matrix of input-outputcoefficients of intermediate products imported by region l and delivered by region k.

In system (2) the trade relationships between region k and l are described by $M_k^l\underline{x}^k$ and $M_l^k\underline{x}^l$, so that fixed import coefficients are assumed. If, however, the export of intermediate products from region k to region l depends on the demand in region l as well as on the share of region k in total national production, the following relationship may be assumed:

$$a_{jl}^{ik} = \rho^{ik}\, a_{jl}^{i} \; , \qquad (3)$$

with a_{jl}^{ik} a coefficient of intermediaries delivered by sector i in region k to sector j in region l, ρ^{ik} a production share of region k in total national production of sector i, and a^i_{jl} a coefficient of intermediate demand by sector j in region l for national commodities produced by sector i.

On the basis of (3) the system for the two regions k and l can easily be rewritten.

A well-known limitation of the input-output approach is that the technical coefficients are assumed to be constant. This implies that each sector has only one average technique at its disposal. In practice, however, very often a choice can be made out of different processes, implying substitution between different inputs. In an interregional framework, also, import substitution between different regions might be considered. Both kinds of substitutions are dealt with later.

A. Energy

The integration of energy in the input-output system is straightforward. In a national input-output system the partitioning of $\underline{x}$, $\underline{f}$ and A leads to (Lesuis et al. 1980a):

$$\begin{bmatrix} \underline{x}^n \\ \\ \underline{x}^e \end{bmatrix} - \begin{bmatrix} A^n_n & A^n_e \\ \\ A^e_n & A^e_e \end{bmatrix} \begin{bmatrix} \underline{x}^n \\ \\ \underline{x}^e \end{bmatrix} = \begin{bmatrix} \underline{y}^n \\ \\ \underline{y}^e \end{bmatrix} \qquad (4)$$

In this system, the indices n and e denote the non-energy and the energy sectors, respectively. The outputs of the energy industries may well be expressed in energy units, and also the energy system itself is representing an input-output system. This extension can be included in systems (2) and (3) in a straightforward manner. In system (4), fixed energy coefficients have been assumed. If different processes of energy production and energy consumption are available, however, energy coefficients may change, whenever changes in relative prices occur. The possibilities of energy substitution are also dealt with later.

B. Pollution

The integration of environmental pollution in the input-output system can be established by defining fixed pollution coefficients with regard to total pollution, (Muller 1973; Nijkamp 1977). In the present study, however, we are especially interested in the amount of pollution caused by energy consumption. Clearly, emission rates are not always the same for all sectors, even if energy consumption is the same. If $v^{i,jr}_p$ is the emission of pollutant p caused by the consumption of energy from sector i in sector j in region r, we obtain: (Lesuis, et al., 1980b)

$$v^{i,jr}_p = q^{i,j}_p \, (a^i_{jr} \, x^{jr} + m^i_{jr}) \qquad (5)$$

In this equation the emission of pollutant p is related to total energy inputs, both originating from the country itself ($a^i_{jr}x^{jr}$, the demand including energy imports from other regions, see (3)) and from abroad (m^i_{jr}) ; $q^{i,j}_p$ is called the emission factor, i.e., the emission rate per unit of consumption of energy source i in sector j. Let $\underline{v}^{er}_p$ and $\underline{v}^{nr}_p$ be row vectors of emissions per unit of product resulting from other sources than energy consumption, caused by energy sectors and non-energy sectors, respectively. Then total pollution becomes (supposing we also pay attention to pollution caused by final demand $\underline{f}$):

$$v^r_p = \underline{i}' \begin{bmatrix} v^{i,jr}_p \end{bmatrix} \underline{i} + \begin{bmatrix} \underline{v}^{nr}_p & \underline{v}^{er}_p \end{bmatrix}' \begin{bmatrix} \underline{x}^{nr} \\ \underline{x}^{er} \end{bmatrix} + \begin{bmatrix} \underline{v}^{fnr}_p & \underline{v}^{fer}_p \end{bmatrix}' \begin{bmatrix} \underline{f}^{nr} \\ \underline{f}^{er} \end{bmatrix} \quad (6)$$

As to the final demand sector, a distinction is made between pollution caused by energy consumption (row vector $\underline{v}^{fer}_p$) and pollution from other sources (row vector $\underline{v}^{fnr}_p$). In the system above, the emission coefficients remain constant. Apart from abatement techniques a reduction in emission may be obtained via substitution processes induced by relative energy prices (as will be discussed hereafter).

2. A TRANSLOG MODEL FOR PRICE EFFECTS AND TECHNOLOGY SHIFTS

The choice of inputs to produce a certain amount of output based on input substitution and technology changes can be treated by incorporating input coefficients in models of producer behavior as endogenous variables, dependent on relative prices (Hudson and Jorgenson 1976). In a multi-regional framework, complexity is increased, however, by including interregional trade possibilities or regional differences between production costs (including environmental factors). Then the competition between regions may cause a specialization of regions and give rise to a specific regional input structure.

In this section, producer behavior will be analyzed on the basis of duality relationships between the production function and the (unit) cost function by using a translog approximation as developed by Christensen et al. (1973). By logarithmic differentiation of the unit cost function with respect to factor prices, a system of relative factor demand relationships is derived, which contains the optimum cost shares necessary to produce a given output at minimum cost. Computation of own price, substitution elasticities and input-output coefficients is straightforward, (Lesuis et al. 1980a). The translog functional form is quite general, since it provides a second-order Taylor expansion of any arbitrary twice differentiable cost function.

The interregional input-output model discussed above is consistent with the following regional translog price possibility frontiers for each sector s and region r:

$$\ln P^{sr} = \ln \alpha_{sr}^{o} + \sum_{ik}\sum \alpha_{sr}^{ik} \ln P^{ik} + \frac{1}{2} \sum_{ikjl}\sum\sum\sum \alpha_{sr}^{ikjl} \ln P^{ik} \ln P^{jl} \qquad (7)$$

where P^{sr} is the output price of sector s in region r, $s = 1,\ldots,I$; $r = 1,\ldots,K+1$, with K the number of regions and $P^{s(K+1)}$ representing prices of imports from abroad.

By differentiating logarithmically with respect to prices and by applying Shephard's lemma, the value shares of the inputs from each sector i and region k in the total inputs in sector s in region r can be derived: (Shepard 1953)

$$S_{sr}^{ik} = \alpha_{sr}^{ik} + \sum_{jl}\sum \alpha_{sr}^{ikjl} \ln P^{jl} \qquad \forall\ s,r,i,k \qquad (8)$$

where the following conditions are valid:

$$\alpha_{sr}^{ikjl} = \alpha_{sr}^{jlik} \qquad \forall\ s,r,i,k,j,l \qquad (9)$$

and:

$$\sum_{ik}\sum \alpha_{sr}^{ik} = 1 \qquad \forall\ s,r \qquad (10)$$

$$\sum_{ik}\sum \alpha_{sr}^{ikjl} = 0 \qquad \forall\ s,r,j,l$$

Equations (8) form a system of regional factor-input demand equations. This system can be conceived of as an interregional input-output model. The derivation of these shares will be discussed later, but first some simplification will be made. We will assume a limited information input-output system, so that (a) the regions of origin k of the inputs remain unspecified, (b) no separate account of the import shares is given, and (c) a regional sector obtains inputs with prices P_r^i within the region r, irrespective of the region of origin k of the inputs. These prices can be considered as region-specific averages f_r^i $(P^{i1} \ldots P^{iK})$ of P^{ik} , $k=1,\ldots K$. This approach leads to a reduction of the number of variables in the specification of the price possibility frontier. A second simplification is to assume that the regional production structure is weakly separable in major categories such as material, energy, capital and labor, (Hudson and Jorgenson 1976). This is consistent with the above mentioned partitioning of the input-output table. As is shown by Fuss, (1977), this implies a two-stage optimization procedure of: (a) the individual inputs of each category and (b) each aggregate input.

This leads to the specification of the following system of equations for the materials sub-model in region r:

$$\ln PM_{sr} = \ln \beta^{o}_{sr} + \sum_{i=1}^{M} \beta^{i}_{sr} \ln P^{i}_{r} + \frac{1}{2} \sum_{i=1}^{M} \sum_{j=1}^{M} \beta^{ij}_{sr} \ln P^{i}_{r} \ln P^{j}_{r} \quad \forall\, s,r \qquad (11)$$

with PM_{sr} an aggregate materials-input price for sector s in region r, P^{i}_{r} an output price of materials or the nonenergy sector i in region r, i=1,...,M with M the number of materials sectors.

Application of the same procedure used in system (7)-(10) leads to a similar specification of the value share equations of individual materials input in total materials input in sector s in region r. The same holds true for the sub-model for energy inputs.

The average input prices for energy and materials, together with exogenous prices of labor and capital services (assumed already to be aggregates) determine regional output prices, which can also be derived from a translog model. Given the resulting share equations and the corresponding restrictions (cf. equations (8)-(10)), all prices can be solved.

The national production of sector s may be allocated to regions according to a similar translog price possibility frontier with its corresponding share equations and restrictions. Such regional sector shares implicitly include interregional trade relationships in the optimization procedure. This causes no difficulties in analyzing energy and pollution problems, as long as traded products have the same pollution characteristics as corresponding non-traded products. In this case, national pollution coefficients may be assumed.

So far, the solution of the model consists of the simultaneous determination of the pattern of economic interactions which result from a given specification of the economic environment. The behavior of the energy sectors is one component of the determination, and the simulated performance of the energy sectors also includes its interrelationships with the rest of the economic system.

Given the projections of the temporal evolution of the vectors of final demand and prices of the primary inputs together with prices of imports from abroad, the projections of a future development for sectors and regions can be made. This development can be judged by a policy-maker to find a balance between divergent objectives, such as employment, energy use, environmental quality and economic growth, which may be conflicting in nature. This will be discussed in the next section.

3. A MULTIREGIONAL MULTIOBJECTIVE POLICY SYSTEM

There is a growing awareness of the existence and relevance of spillover effects, both between economic subjects and between regions. A simultaneous consideration of all relevant policy objectives (implying a multidimensional objective profile) and of all relevant regional decision units (implying a multiregional profile) complicates the traditional decision and programming methods. Therefore, a new formal approach based on a generalized multiobjective programming framework has to be devised, such that the interdependencies among the various elements of the policy structure are reflected.

The conflict between regions emerges from the existence of spillover effects in an open spatial system (for example, interregional input-output linkages, diffusion of pollution, transportation), while the conflict between objective functions emerges from intraregional interactions such that the achievement of a high value of the one objective involves a low value of a competing objective. The latter type of conflict is studied in the field of multiobjective programming. (Cohon 1978; Delft and Nijkamp 1977; Keeney and Raiffa 1976; Nijkamp 1979).

The existence of competing regions can formally be described by the same multiobjective approach. Suppose that the integrated energy-economic-environmental structure of region 1 can be described by means of the following model:

$$\underline{x}_1 = f\,(\underline{x}_1\ ,\ \underline{x}_2\ ,\ \underline{e}_1) \qquad (12)$$

where $\underline{x}_r$ is a set of relevant variables for region r, r = 1,2 (for example, sectoral production levels, employment, emission of pollution, energy consumption, etc.), and where $\underline{e}_1$ represents a set of exogenous variables.

Clearly, an analogous model can be constructed for region 2. The interregional input-output model discussed above can be regarded as a further specification of such a model.

In addition to the structural relationships incorporated in (Nijkamp 1977), a set of regional side-conditions (technical, economic, environmental, institutional) may be assumed. Together with Nijkamp the feasible area of $\underline{x}_1$ may be represented by K_1:

$$\underline{x}_1 \in K_1 \qquad (13)$$

Then the following multiregional multiobjective programming problem for the spatial system as a whole can be assumed:

$$\begin{array}{ll} \max\ l_1\,(\underline{x}_1) & \max\ l_2\,(\underline{x}_2) \\ \max\ s_1\,(\underline{x}_1) & \max\ s_2\,(\underline{x}_2) \\ \max\ u_1\,(\underline{x}_1) & \max\ u_2\,(\underline{x}_2) \end{array} \qquad (14)$$

$$\text{subject to:}\quad \underline{x}_1 \in K_1\ ,\ \underline{x}_2 \in K_2.$$

In the present paper, the assumption will be made that the spatial system as a whole aims at achieving a maximum value for the three successive objective functions, while next on the basis of a compromise solution the regional authorities aim at achieving the most favorable outcome for the region at hand. This approach will be based on an interactive learning procedure, so that the centrally coordinated decisions have to take account of the regional priorities, while on the other hand the regional options are co-determined by national priorities (see for a formal exposition of multilevel multiobjective programming Nijkamp and Rietveld 1981).

The interactive approach used in the present article is based on a series of successive steps. These steps will briefly be discussed:

(a) The first step implies a (centralized) optimization of the three regionally aggregated objective functions:

$$\begin{aligned} &\max\ l = l_1 + l_2 \\ &\max\ s = s_1 + s_2 \\ &\max\ u = u_1 + u_2 \\ &\text{subject to } \underline{x} \in K \end{aligned} \qquad (15)$$

The optimal solutions of each separate optimization of the three objective functions of (15) are denoted by (l_1^o, l_2^o), (s_1^o, s_2^o), and (u_1^o, u_1^o), while the optimal values of the corresponding argument variables are denoted by (x_1^l, x_2^l), (x_1^s, x_2^s) and (x_1^u, x_2^u) respectively

(b) The compromise solution between the three aggregate objective functions can be found by constructing a so-called pay-off matrix P (Fandel 1972; Nijkamp and Rietveld 1976). Such a pay-off matrix reflects the losses in a certain objective, when a competing objective is maximized. A compromise between such conflicting objectives can be achieved by means of equilibrium notions from game theory. There are several ways to identify such an equilibrium point from a pay-off matrix P.

The elements on the main diagonal of P represent the so-called ideal points, i.e., the solutions of model (15). The off-diagonal elements represent the values of a certain objective function, when the optimal argument variables corresponding to the maximum of another objective function from model (15) are substituted into this function. Consequently, P shows the sacrifices in other objectives when a certain extreme option (the maximum of only one objective function) is achieved. Clearly, any meaningful compromise between these extreme options should fall in the range of the row minima and row maxima of P.

It should be noted that the pay-off matrix can be divided into regional components, so that the diagonal blocks of P reflect again the intra-regional conflicts, and the off-diagonal blocks the inter-regional conflicts. Then the diagonal elements in the off-diagonal blocks are related to conflicts between the same set of objective functions in different regions.

There are various ways to calculate a compromise solution from the pay-off matrix P, for example, by calculating the vector of weights $\underline{\lambda}$ of the successive objective functions for which all extreme solutions of P are valued equally, (Fandel 1972).

(c) The regional compromise solutions (l_1^*, s_1^*, u_1^*) and (l_2^*, s_2^*, u_2^*) are provided to regions 1 and 2 as a frame reference. These regions have to judge whether or not they are satisfied with the initial compromise solution. When a region deems a certain compormise unacceptable, this compromise solution should reach a higher value. The set of objective functions which are regarded as unsatisfactory are denoted by S_1 and S_2, respectively. Thus, step (c) can be regarded as the specification of decentralized priorities regarding the achievement of objectives.

(d) The decentralized priorities of step (c) can be incorporated as constraints in the next phase of the interactive program. These constraints can be formalized as:

$$\begin{array}{ll} l_1 > l_1^* & l_2 \geq l_2^* \\ s_1 > s_1^* & s_2 \geq s_2^* \\ u_1 > u_1^* , & u_2 \geq u_2^* , \\ \text{if } l_1, s_1 \text{ or } u_1 \in S_1 & \text{if } l_2, s_2 \text{ or } u_2 \in S_2 \end{array} \qquad (16)$$

These constraints can now be included in the first step of the second run of the interactive process in which the central policy solution is calculated [see also (15)]:

$$\begin{aligned} &\max l = l_1 + l_2 \\ &\max s = s_1 + s_2 \\ &\max u = u_1 + u_2 \qquad (17) \\ &\text{subject to: } \underline{x} \in K \\ &\qquad \text{and condition (16)} \end{aligned}$$

Then the procedure can be repeated, until a convergent solution is obtained (see for a convergence proof (Fandel 1972 and Rietveld 1980). This ultimate compromise solution can be regarded as an equilibrium solution between the diverging options of conflicting objectives and conflicting regional interests.

4. CONCLUSION

The approach described in the previous sections has been applied to an interregional input-output model for the Netherlands, (Lesuis, et al. 1980b). The provisional results obtained so far demonstrate the operational character of the abovementioned analysis. Both the translog model and the multiobjective model based on some meaningful policy scenarios provide meaningful results.

NOTES:

[1] In the article we adopt the convention of denoting regions and sectors from which products originate (as outputs) by superscripts, and regions and sectors which use products (as inputs) by subscripts.

BIBLIOGRAPHY

Christensen, L.R., D.W. Jorgenson and L.J. Lau, 'Transcendental Logarithmic Production Frontiers,' Review of Economics and Statistics, Vol. 55, pp. 28-46, (1973).

Cohon, J.L., Multiobjective Programming and Planning(Academic Press, New York), (1978).

Delft, A. van, and P. Nijkamp, Multicriteria analysis and Regional Decision-Making, (Martinus Nijhoff, Boston/The Hague), (1977).

Fandel, G., Optimale Entscheidung bei mehrfacher Zielsetzung, (Springer, Berlin), (1972).

Fuss, M.A., 'The Demand for Energy in Canadian Manufacturing', Journal of Econometrics, Vol. 5, pp. 86-116, (1977).

Hudson, E.A. and D. W. Jorgenson, 'Tax Policy and Energy Conservation', in: D.W. Jorgenson (ed.), Econometric Studies of U.S. Energy Policy, (North-Holland Publ. Co., Amsterdam), pp. 9-94, (1976).

Keeney, R.L. and H. Raiffa, Decision Analysis with Multiple Conflicting Objectives, Wiley, New York), (1976).

Lesuis, P.J.J., F. Muller and P. Nijkamp, 'Operational Methods for Strategic Environmental and Energy Policies', in: T.R. Lakshmanan and P. Nijkamp (eds.), Environmental-Energy-Economic Interactions, (Martinus Nijhoff, Boston/The Hague), pp. 40-73, (1980a).

Lesuis, P.J.J., F. Muller and P. Nijkamp, 'An Interregional Policy Model for Energy-Economic Environmental Interactions', Regional Science and Urban Economics, Vol. 10, nr. 3, pp. 343-370, (1980b).

Muller, F., 'An Operational Mathematical Programming Model for the Planning of Economic Activities in Relation to the Environment', Socio-Economic Planning Sciences, Vol. 7, pp. 123-38, (1973).

Muller, F., Energy and Environment in Interregional Input-Output Models, (Martinus Nijhoff, London/The Hague/Boston), (1979).

Nijkamp, P., Theory and Application of Environmental Economics, (North-Holland Publ. Co., Amsterdam), (1977).

Nijkamp, P., Multidimensional Spatial Data and Decision Analysis, (Wiley, New York), (1979).

Nijkamp, P. and P. Rietveld, 'Multiobjective Programming Models', Regional Science and Urban Economics, Vol. 7, pp. 253-74, (1976).

Nijkamp, P. and P. Rietveld, 'Multiobjective Multilevel Programming in a Multiregional System', in P. Nijkamp and J. Spronk (eds.), Multicriteria Analysis: Practical Methods, (Gower Publ. Co., London), Forthcoming, (1981).

Rietveld, P., Multiple Objective Decision-Making and Regional Planning, (North-Holland Publ. Co., Amsterdam, (1980).

Shephard, R.W., Cost and Production Functions, (Princeton University Press, Princeton), (1953).

7 National-Regional Interdependencies in Integrated Economic-Environmental Energy Models

W. HAFKAMP AND P. NIJKAMP

1. INTRODUCTION

The call for abatement of air pollution has been rising during the last decade. On the one hand, this is caused by the recently growing awareness of the dangers of air pollutants for human health and the biotic environment in general; for instance, the negative effects of acid rains on the natural environment have been shown to be tremendous. On the other hand, this is caused by substitution processes in the energy sector; for instance, availability of natural gas in the Netherlands urges energy users to switch back to oil or even to coal. The Maximum Acceptable Concentrations of air pollutants have been set increasingly lower over the last decade. In that case, any further increase of emissions of air pollutants - due to a forced shift to more polluting fuel types and/or a general increase in the demand for energy - is a serious problem.

In the Netherlands, authorities are increasingly becoming aware of the fact that the emission of among others, sulphur dioxide and nitrogen oxides has to be reduced substantially. Therefore, the Dutch government has recently started working on a series of concrete environmental quality measures. The seriousness of environmental problems has led the authorities to edict measures in a policy atmosphere, even before a thorough study of the economic consequences of these measures has been completed. In this paper, an attempt at providing a background study will be presented.

The present paper will provide a framework for estimating the effects of anti-pollution measures and regional economic policy measures on the emission and immission of some important categories of air pollution, regional income, and regional and sectoral employment. The growing scarcity of important fuel types such as oil and natural gas will also be an essential element in this paper.

The method of analysis being used here has been described in Hafkamp and Nijkamp (1979a) and Hafkamp (1979). This method consists of two phases:

1. The construction of a conceptual model on the basis of problem definition, experience and scientific background.

2. The deduction of an operational model from the conceptual model and available data.

The necessity of constructing a conceptual model is argued in Hafkamp and Nijkamp (1979a). A conceptual model is constructed there, while a brief outline of an operational model is also given there. Here we will elaborate on the operational model. In this paper we shall present the design and the structure of a so-called Triple-Layer-Model (TLM). Next, we shall deal with the separate submodels and their interrelations in a more detailed way. Some attention is also devoted to operationalizing the model and simulating a decision-making process with various policy options. Some conclusions and future research lines will be given in section 5.

2. DESIGN OF A TRIPLE-LAYER-MODEL

In this section, a model of a society with production, employment and pollution will be presented. This will be done by projecting reality on three surfaces associated with three interrelated submodels, respectively:

E: a national-regional economic model
L: a regional, sectoral employment model
P: a pollution model, against the background of energy availability.

Together the three submodels form the Triple-Layer-Model (TLM). The model is not entirely new in all its components. Part of the economic model is based on the so-called SECMON-model of the Dutch economy by Driehuis (1979). The SECMON-model is a large simulation model of the Dutch economy (including four production sectors). We shall regionalize the model into five regions and disaggregate it into 11 sectors. This makes the model itself a multi-level model in which national-regional interdependencies reflect the relations between the two levels of the model. Models like this have been developed in France, Belgium and The Netherlands. For France, the REGINA-model (see Courbis (1979)) describes a three-level system which aims at optimizing regional employment conditions by a policy of regional investment planning. For the Netherlands, the REM-model (see Van Delft and Van Hamel (1978)) does the same for a two-level structure in a less detailed manner. For Belgium, there is the RENA-model (see De Corel, Thys-Clement and Van Rompuy (1973)). A detailed methodological comparison between these models is made by Hordijk and Nijkamp (1980), where also a set of criteria is formulated, which a national-regional model has to fulfill.

The labour market model describes employment (supply and demand) in the regions and sectors. The demand for labour is analyzed through the production structure; gross production, capacity and capacity use in capital-intensive sectors as well as import substitution are important elements here. The supply of labour is analyzed through wages, prices and income developments and further with given demographic data. This submodel is mainly related to demand for labour, because anti-pollution measures and stimulation of employment are often regarded as contradictory options (though this is not necessarily the case).

The pollution model contains a description of emission and diffusion of (air) pollutants. Introduction of anti-pollution technology takes place in this model. It should be noted however, that we cannot limit ourselves to only pollution aspects as such. A major part of the air pollution is caused by combustion of so-called fossil fuels. Therefore, it is necessary to take into account the availability of several types of energy, its use in the production and consumption sector as well as its environmental repercussions.

The three block diagrams in Figure 1 can best be seen as parallel planes situated above each other; inter-layer relations are indicated by vertical arrows, whereas intra-layer relations are indicated by horizontal arrows.

In the next sections we present the design of the TLM. Variables and relations of Figure 1 are further specified here. Every submodel is presented seperately. Firstly, the variables and relations within the submodels are treated, while next the interdependencies between the submodels will be dealt with.

3. THE REGIONAL-NATIONAL ECONOMIC MODEL

The submodel is based on the regional input-output tables for the 11 Dutch provinces for 1970, published by the Central Bureau of Statistics (CBS). Behavioral equations and the blocks associated with the government and social insurances stem from the SECMON-model and are regionalized (if necessary).

The production block is based on the above mentioned input-output tables. For this purpose the 11 Dutch provinces are aggregated to 5 regions. The sectors are aggregated to 11 sectors. Figures for 1970 are updated for the year 1975 (in a later phase the actual input-output tables for 1975 will be published by the Dutch CBS). The following equations will be used to calculate final demand at a sectoral and regional level:

- Consumption (private): The value of total private consumption is calculated for every region via regional wage and non-wage income. Regional expenditure patterns are assumed to be equal.

- Investments (by firms): The level of the investments by firms is determined by various factors such as wages, non-wage income, prices, capacity variables and monetary variables. A distinction is made between investments in buildings and investments in equipment. For agriculture and the capital-extensive service sectors, it is assumed that all investments are delivered by the construction sector of the same region. In the manufacturing sectors and the capital-intensive service sectors, investments are delivered by the construction sector according to a constant ratio.

- Government expenditure:
  Material government consumption is mainly considered as exogenous (e.g., military expenditures, and other categories of government consumption). An example of an endogenous material government consumption is the public expenditure which strongly depends on the number of government employees.
  Government investments are composed of two categories. Investments in road construction and water control are assumed to be endogenous (depending upon investments in housing and the

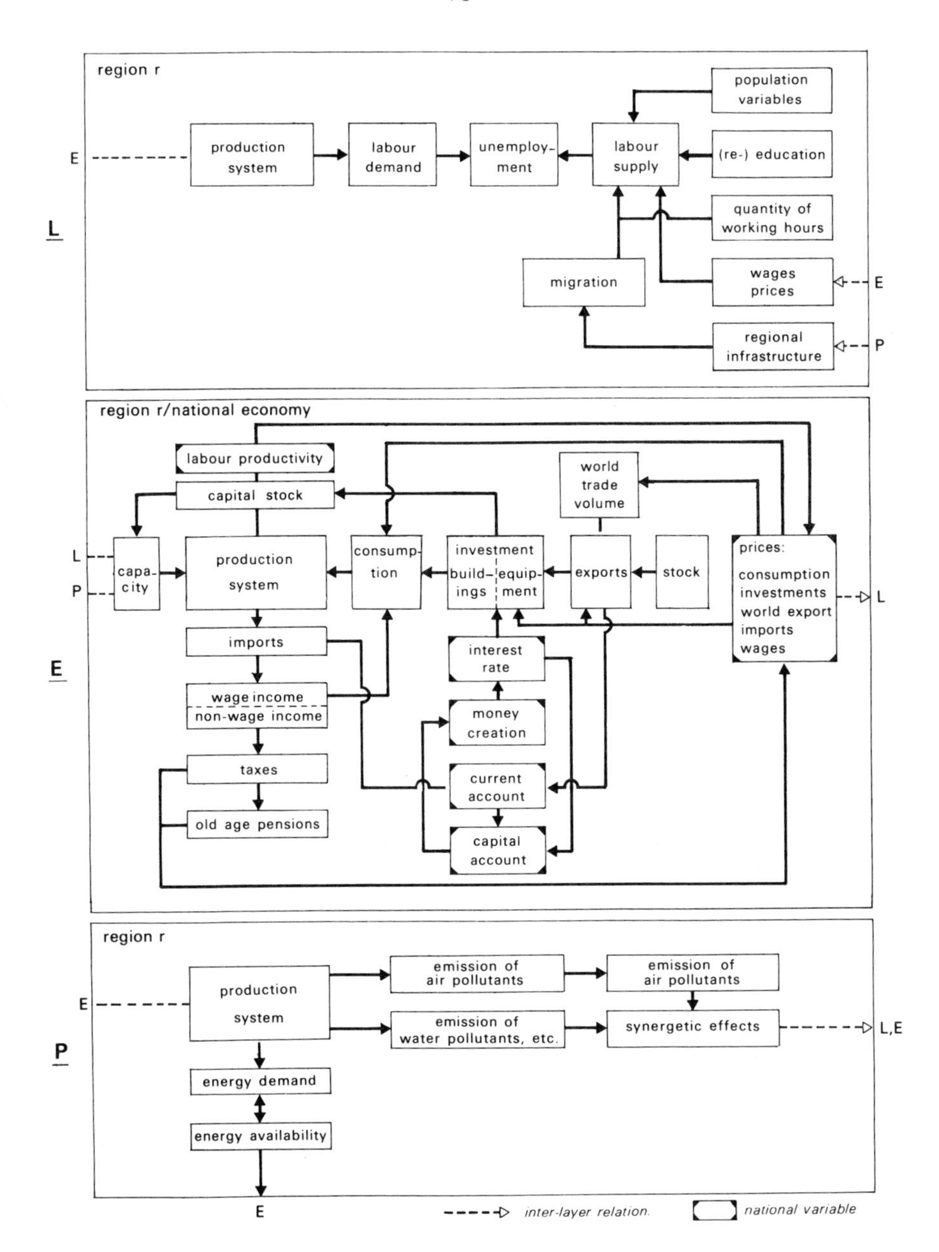

Figure 1. Structure of a Triple-Layer-Model.

interest rate), while other investments are assumed to be exogenous.

- Exports: Detailed information on interregional imports and exports by sector is not available. Only national exports - the most important category - are analyzed in the model. They are assumed to depend upon the volume of world trade and the relative prices of Dutch exports.

- Imports: A distinction is made between: (1) final products and (2) raw material/manufacturing inputs. Category 2 is subdivided into: (a) competitive and (b) non-competitive imports. The volume of category 2 imports depends on the level of gross production, while the volume of category 2a also depends on the relative price level.

Production capacity is calculated for the manufacturing industries and for the capital-intensive service sectors. Growth of production capacity is determined by the investment rate in equipment.

Wages and prices are calculated at a sectoral level, but not at a regional level. Regional differences of wages and prices are so small that they may be neglected. Contract wages in the manufacturing industries are established in negotiations between employers and employees. Influencing factors are: expected consumer prices, specific sectoral circumstances and compensations for tax rise and social insurance premiums. Producers' prices are determined by other producers' prices, wage costs per unit of product, prices of imported inputs, capacity rate and capital costs.

Taxes consist of endogenous tax variables such as indirect taxes minus subsidies (e.g., TVA), company taxes, income taxes and exogenous tax variables such as revenues from natural gas resources and other non-tax incomes. Non-material government expenditures are wages, interest payments and income payments to individuals and firms.

The monetary sector is described at a national level. The most important variables here are money creation (by the government and by firms) and the interest rate. These variables have an influence on sectoral investments and production capacities.

In this economic submodel, there are bottom-up relations as well as top-down relations. Bottom-up relations are especially relevant for consumption, prices, wages, investments, exports and imports. Top-down relations relate in particular to government expenditures and the monetary sector, but also to the total level of private investments. Variables like gross production, imports, exports and prices of several types of goods are also elements in the other two sub-models.

4. THE LABOUR MARKET MODEL

Labour demand in regions and sectors is analyzed in the labour market model. For the time being, we consider the supply side given. Demand for labour is assumed to depend mainly on gross production in regions and sectors. Besides, there are other important factors such as quantity of working hours, capacity rate and labour-saving technological progress.

5. THE POLLUTION MODEL

This submodel deals with three phenomena:

1. Emission of air pollutants caused by:

   a. combustion of fossil fuels
   b. process emissions, etc.

2. Concentration of air pollutants

3. Reduction of emission by:

   a. energy saving, selective growth, etc.
   b. alternative choices of energy sources
   c. anti-pollution technology

Pollution of water and soil is not taken into account here. Also, no attention is paid to the phenomenon of synergetic effects. This means that any anti-air-pollution program has to be accompanied by other programs through which the natural environment is protected on other aspects than air pollution. In general, a substitution of one type of pollution by another one has to be prevented in pollution management programs. Not all types of air pollution will be studied in this section. We limit ourselves to three important categories: sulphur dioxide, nitrogen oxides and dust particles.

Emission of air pollutants is subdivided into two categories (see 1a and 1b above). Process emissions make up less than 10% of the weight of total emissions of air pollution. Therefore, we deal only with combustion emissions. Gross production and production techniques determine the quantity of various kinds of fuel to be used. Then the emission of air pollutants can be calculated through a vector of emission coefficients (see Institute for Environmental Studies (1978)). Special points of interest are the integration of production and demand for electricity as well as pollution by traffic.

Diffusion of air pollution is important, because over the last decade the maximum acceptable concentration of SO_2 was more and more surpassed in several places in the Netherlands. It has to be added that also emissions from German and Belgian industrial areas influence the concentration of air pollutants in some parts of the Netherlands, pending on the weather conditions. The diffusion of air pollutants depends on many factors (inter alia place, time, quantity and height of emission and weather conditions). Diffusion models can be constructed analytically (see among others Muller (1979) and Coupe (1975)) or empirically. If an empirical approach is chosen, one has to obtain emission data over a certain period at several sources in an area as well as imission data from a number of places in the whole area where the emissions influence the concentration of air pollutants. For the moment there is no such model for the Netherlands. We use in our study a diffusion model developed by Coupe (1975).

In the beginning of this section, anti-pollution measures have been subdivided into three categories. Here it becomes clear that there is an important correspondence between the energy problem and the pollution problem. Much research has already been done in the field of purification of flue glasses and clean combustion methods (see among others Pearse and Seaman (1975) and Zinn and Lesso (1978)). In our submodel a number of abatement techniques is introduced to reduce emission to a fraction of its original level (10% to 90%). The necessary investments and other costs are integrated into the economic system via an additional sector. The way these programs are financed will be treated in the next section.

In the beginning of this section, anti-pollution measures have been subdivided into three categories. Here it becomes clear that there is an important correspondence between the energy problem and the pollution problem. Much research has already been done in the field of purification of fluegasses and clean combustion methods (see among others Pearse and Seaman (1975) and Zinn and Lesso (1978)). In our submodel a number of abatement techniques is introduced to reduce emission to a fraction of its original level (10Δ to 90Δ). The necessary investments and other costs are integrated into the economic system via an additional sector. The way these programs are financed will be treated in the next section.

The choice of the energy source also has an important influence on the emission of air pollutants. For example: SO_2 emissions in the Netherlands decreased drastically after a large-scale introduction of natural gas, but since a switch back to coal or oil may take place, a drastic increase may occur. Especially the shift of electricity producers from natural gas to oil, coal or nuclear energy and the further exploration and introduction of alternative energy sources (solar energy, wind, etc.) are of great importance to environmental quality.

Besides these measures, there are other means to influence air pollution via energy use. Selective growth is a good example of this, where the growth in sectors which cause little pollution due to a small energy use is favoured. Another important aspect in western countries is the effect of a decreasing population, which also leads to a relatively decreasing demand for energy. Also, there is a call for an "energy tax" in the Netherlands, which may cause a general decrease of the demand for energy over all sectors. Any type of energy conservation (insulation, etc.) leads to a decreasing energy use and thereby to a decreasing emission of air pollution.

6. SIMULATION, DATA BASE AND DECISION-MAKING

The model as a whole is a simulation model. It does not provide us with an adequate explanation of society and human behaviour. The model is not appropriate to make reliable predictions for the nearby future. The data base for this model is relatively small; clearly, information should be available at a much more disaggregated level. A number of SECMON equations have a low R^2, while not all coefficients differ significantly from zero (at a .95 level). Simulations with the model are characterized more by an optimization approach than by a scenario approach, although there are certain elements of a scenario approach present in the model. Within an optimization approach, the value of one or more goal variables has to be optimized, given a set of systems constraints and policy options. Within a scenario approach the consequences of alternative policy options for the system are studied. In our case the goal variables are (regional) income, environmental quality and labour opportunities. In this study a multiple-objective, multi-decisionmaker (MOMD) problem has to be solved. Elements of a scenario approach are here also present, because a decision-making procedure can start from different viewpoints. Especially, financing anti-pollution measures is an important issue here. A broader discussion of tax/subsidy schemes and redistribution principles (polluters pay, pollutee pay, etc.) is given in Hafkamp and Nijkamp (1979a, 1979b). The MOMD problem at hand contains partly conflicting interests. Earlier studies have shown that there is rather a conflict between income and environmental quality than a conflict between energy and environmental quality (see Nijkamp (1980)). This is clear, because the improvement of environmental quality can often be seen as an economic (public) good.

6. CONCLUSION

The conclusions of this paper are, in a way, preliminary, because the actual construction of the model has not entirely been completed (see further Hafkamp and Nijkamp, (1981)). The model, however, may be a valuable contribution to the current scenario models of energy economic systems and national-regional models. It is an ambitious attempt to construct an integrated regional-economic-environmental-energy model, which may include both an optimization structure and a scenario structure. In this respect this model can be used as an important tool for comprehensive policy analysis.

BIBLIOGRAPHY

Corel, L. de, F. Thys-Clement and P. van Rompuy, RENA - Een Econometrisch Model voor het Plan 1976-1980, Planbureau Brussel, (1973).

Coupe, B., Regional Economic Structure and Environmental Pollution, Stenfert Kroese, Leiden, (1977).

Courbis, R., 'The Regina Model - A Regional-National Model for French Planning,' Regional Science and Urban Economics, Vol. 9, pp. 117-139, (1979).

Delft, A. van and B. van Hamel, 'Operationeel Onderzoek en Regionaal Beleid', in Regionaal Beleid, P. Nijkamp and C. Verhage (eds.), Stenfert Kroese, Leiden/Antwerpen, pp. 123-156, (1978).

Driehuis, W., 'Een Sectoraal Model t.b.v. de Analyse van de Nederlandse Economie', University of Amsterdam, (mimeographerd), (1979).

Hafkamp, W., Elements of Science Policy in Environmental Economics, University of Amsterdam (mimeographed), (1979).

Hafkamp, W. and P. Nijkamp, 'Dilemma's in Environmental Economics: Environmental Models Revisited,' in Canadian Journal of Regional Science, Vol. 11, No. 2, pp. 1-21, Institute of Public Affairs, Halifax, Canada, (1979a).

Hafkamp, W. and P. Nijkamp, 'An Integrated Interregional Model for Economic-Environmental Policy,' in Economic-Environmental-Energy Interactions, T.R. Lakshmanan and P. Nijkamp, Eds.), Martinus Nijhoff, Boston, pp. 149-171, (1979b).

Hafkamp, W. and P. Nijkamp, 'Towards an Integrated National-Regional Environmental-Economic Model,' paper presented at the IFIP Conference on Environmental System Analysis and Management, Rome, September 1981.

Hordijk, L. and P. Nijkamp, Integrated Approaches to Regional Development Models, Research Memorandum 1980 - 4, Dept. of Economics, Free University, Amsterdam, (1980).

Institute for Environmental Studies, Milieuverontreiniging en Productiestructuur in Nederland, Free University, Amsterdam, (1978).

Muller, F., Energy and Environment in Interregional Input-Output-Models, Martinus Nijhoff, Boston/The Hague/London, (1979).

Nijkamp, P., Environmental Policy Analysis, John Wiley a Sons, Chichester/New York, (1980).

Pearse, J. and M. Seaman, Economic Consequences of NO_x and Odour Abatement, Interim Report for the Ministry of Health and Environmental Control, The Hague, (1975).

Zinn, C.D. and W. G. Lesso, 'Putting the Economics in Best Available Control Technology or Pollutants in the Air and Dollars to the Wind,' Interfaces, Vol. 9, No. 1, (1978).

8 A Programming Approach as a Design for Economic Development Policy

R. BANNINK, C. BROEKHOF AND P. NIJKAMP

1. INTRODUCTION

The formulation of economic policy needs information from various fields. Without denying the importance of other auxiliary disciplines, we will concentrate ourselves mainly on an economic analysis in this preparative study. Economic policy analyses are usually based on econometric models describing equilibrium growth (e.g., based on vintage production structures). An analysis of the changing input values of these models provides the expert with more insight into reactions from medium- to long-term forces on these changes. By trial and error procedures one may try to identify those changes which are in close agreement with the policy-makers evaluation.

As each major problem requires its own specific model (e.g., in terms of employment, energy, pollution, etc.), it is an important responsibility for policy-makers to translate the insights regarding key elements of problems - obtained by studying such models - into a consistent policy.

In the present paper an alternative approach based on programming techniques will be presented.

The reasons why we prefer models using these techniques are:

- they can start from any historical set of data independently of the existence of an equilibrium or a disequilibrium situation on the commodity, labour or money market.
- they can easily - at least in a conceptual sense - integrate several relevant aspects of policy-making such as a trade-off analysis.
- they can describe both a disequilibrium and an equilibrium path.
- they focus much attention on the definition and relevance of policy objective functions
- they allow the description of a multiple goal-setting, caused by the participation of different groups in policy-making.
- they focus the attention on the importance of specific

restrictions in relation to the simultaneous effects of all other restrictions.

This paper is a new one in a series of reports on this project, which started in 1975 with a paper at the ISI Conference in Warsaw (September 1975). The structure of the paper and the models described therein reflect our efforts to describe the essence of our thoughts, rather than their historical development, statistical difficulties on computational results. In section 2 we will describe the core of the economic structure. In section 3 we will introduce pollution and energy. In section 4 the balance of payments will be considered. In section 5 the public sector will be described. In section 6 we reconsider policy objective functions and conclude this paper.

2. THE CORE OF THE MODEL

The core of the model is based on the idea that production is realized in production sectors by means of a combination of primary inputs, labour and capital goods; this structure can be described by an extended Input-Output-model. This model is called extended, because traditional I-O-models omit capacity limits.

A second aspect of the core model is the circularity: the consumption sector earns its income from these production activities and decides to spend its income to consumption or savings.

A third aspect is the dynamic structure reflecting the autonomous progress of technology, which can be included in the economy by investments.

The final aspect is the set of policy objective functions which are related among others to decisions to produce and to invest in the planning period.

2.1. PRODUCTION FUNCTIONS

For production sector i, there is a production function in which the production depends *inter alia* on the year of investment of each capacity used. This structure is described by the following equations:

$$QP(i,t) = \sum_{\tau=t-\ell}^{t-1} QP(i,t,\tau) \qquad (1)$$

where production of sector i in year t equals the sum of the production of that sector in period t produced by using the productive investments installed in year . Technical life time is defined by . Construction of capacity is assumed to take 1 year.

$$QP(i,t,\tau) \leq \frac{FINV(i,\tau)}{PINV(\tau) * \kappa(i,\tau)} \qquad (2)$$

Production capacity is smaller than or equal to the value of investment divided by its price and marginal capital-output ratio.

$$QLAB(s,i,t,\tau) = \alpha(s,i,\tau)*QP(i,t,\tau) \qquad (3)$$

Labour is subdivided into categories, denoted by the index s. Each category has an investment-specific productivity, denoted by its reciprocal value $\alpha(s,i,\tau)$. This parameter has an exogeneous temporal evolution.

$$\alpha(s,i,\tau) = A_{s,i}[\exp B_{s,i} \cdot \tau] \tag{4}$$

where $A_{s,i}$ and $B_{s,i}$ are input parameters reflecting the progress in technological know-how.

$$\begin{aligned} FREV(i,t,\tau) = {} & P(i,t)*QP(i,t,\tau) \\ & - \sum_{s} W(s,t)*QLAB(s,i,t,\tau) \\ & - \sum_{j} P(j,t)*\beta_{ji}*QP(i,t,\tau) \geq 0 \end{aligned} \tag{5}$$

The gross profit from employing a time-specific capacity for production has to be positive. New variables in this equation are:

$P(i,t)$ denoting the prices of products from branch i in year t;

$W(s,t)$ denoting the wage rate for labour capacity s in year t;

β_{ji} denoting the input-output coefficients of intermediate production of sector j needed to produce one unit of production in sector i.

Contrary to the commonly used models, the price and wage variables are given input parameters in order to maintain a linear model. However, a reaction of labour income on profits is taken into consideration in equation (7).

2.2. INVESTMENTS

At the outset of the planning period, capacity is given for each sector of industry by the historical investment series. Since the model is solved for all periods within the planning period simultaneously, the future demand for each sector of industry is known (see equation 11).

The amount of investments, needed to adjust capacity to this demand, has to satisfy a pay-back criterion, commonly used in industry:

$$FINV(i,t) \leq \sum_{\tau=t+1}^{t+T_i} FREV(i,\tau,t) \tag{6}$$

where T_i = (critical) value of the pay-back period in branch i.

Adjustment of production capacity to demand is not the only reason for investments: as soon as profit is an element in the objective function, investments can be made to raise profitability by using relatively more recent equipment, thus raising labour productivity and thereby reducing the wage costs.

Common economic theory describes a wage rate reaction to this policy, but wage rates and prices have to be treated as exogenous variables in this model to keep its linear structure.

We employ the following equation to reflect the wage reaction on a rising profitability:

$$FSUR(i,t) = \sum_{t'=t-T_i}^{t-1} \lambda(t,t') \sum_{\tau=t'+1}^{t'+T_i} FREV(i,\tau,t')-FINV(i,t') \quad (7)$$

The expression in brackets is the slack in equation (6), the surplus profit in relation to the minimum profit level needed to invest. The $\lambda(t,t')$'s are a series of distributed-lag parameters, only depending on the difference between t and t' and adding up to a value equal or less than unity. The amount FSUR(i,t) is then the total effect in period t of all these shares in surplus profits realized by investments before t. This quantity is as well a part of labour income as normal wages are.

By this formulation we avoid to turn down hardly profitable sectors of industry into submarginal profitable ones. Only the timing of earning the surplus profit and paying the related wage quota can differ slightly.

A third restriction on investments is that the financial means have to be available. These financial means can be provided by the initial capital, enlarged during the planning period by retained profits. These funds can - for each sector i - be enlarged by borrowing savings from consumers, but in that case a maximum ratio between capital growth from retained profits and from new savings has to be taken into account (see equation (9)):

$$\sum_{\tau=t_o}^{t} FINV(i,\tau) \geq CAP(i,t_o) + \sum_{\tau=t_o}^{t-1} \{\sum_{\tau} FREV(i, \ , \ ')-FSUR(i, \ )\} + \sum_{\tau=t_o}^{t-1} SAV(i,\tau) \quad (8)$$

$$\sum_{\tau=t_o}^{t} SAV(i,\tau) \leq \phi(i) \sum_{\tau=t_o}^{t} \{\sum_{\tau} FREV(i,\tau,\tau')-FSUR(i,\tau)\} \quad (9)$$

$$\sum_{i} SAV(i,t) \leq FSAV(t), \quad (10)$$

where FSAV(t) are total consumer savings in period t.

Some final remarks have to be made before concluding this subsection.

In equation (6) a correction has to be made for the last years of the planning period to avoid disturbing horizon effects. We have chosen a factor $\frac{T_i}{T-t}$ for $T - t < T_i$, where T is the last period before the planning horizon. In that case the summation limit is T.

In equation (8) the summations in the right-hand side are only relevant for t-values larger than t_0, the first planning period.

2.3. DEMAND

The final demand for consumption goods is given by the Keynesian consumption function:

$$FDEM(i,t) = CO(i) + MC(i,t) * FINC(t), \qquad (11)$$

where MC(i,t) is the marginal consumption rate and where $\sum_i MC(i,t) < 1$ is the aggregate marginal consumption rate.

Income is defined by:

$$FINC(t) = \sum_{s,i,\tau} W(s,t)*QLAB(s,i,t,\tau) + \sum_i FSUR(i,t) \qquad (12)$$

The final demand for capital goods is given by adding up all planned investments FINV(i,t) over i. The result of this summation is supposed to be stored in FDEM(i,t) for i referring to capital producing industry, where CO(i) and MC(i,t) for that industry are zero.

Now we reach a second deviation from traditional equilibrium models, which is even more fundamental than the first one which has led us to the definition of FSUR(i,t). The restriction on production by capacity and profitability can cause a shortage in supply as compared with demand. Given the exogenously determined wage rates and product prices, the producers can even decide to stop production. Another reason can be found in the restrictions on investment funds leading to undercapacity and consequently to a shortage in supply. The non-negative gap between supply and demand is supposed to be delivered by external suppliers. The size of this external supply is a signal for the user of this model to consider whether changes in the parameters are needed or not.

$$\sum_{\tau} QP(i,t,\tau)*P(i,t) + FIMP(i,t) = \sum_j P(i,t)*'\beta_{ij}*\sum_{\tau} QP(j,t,\tau) + FDEM(i,t) \qquad (13)$$

2.4. LABOUR MARKET

In this model the labour market is not a market in the ordinary sense, where equilibrium is found between supply and demand, but a reservoir. For each category of labour, no more can be used than is available, while available labour is an exogenous variable.

$$\sum_{i,\tau} QLAB(s,i,t,\tau) \leq LAB(s,t) \qquad (14)$$

This formulation supposes a perfect mobility between sectors of industry within each labour category and perfect immobility between categories.

2.5. CAPITAL MARKET

There is also for the capital market just a minor function conceptualized in this model. For the productive use of savings we defined already equation (10). Here we have to define the amount of savings:

$$FINC(t) + FCI(t) = \sum_{i \varepsilon C} FDEM(i,t) + FSAV(t) \qquad (15)$$

This equation defines savings as the balance between income and consumption. There are two variables concerned with savings: FSAV for positive savings and FCI for negative savings; their product has to be zero. Technically, we take account of the latter remark by giving FCI an unfavourable coefficient in the goal function, which exceeds the sensible range of shadow prices of FSAV. The quantity FCI is external just as FIMP.

Of course, the model could be enlarged by interest from savings, in which case an exogenous price variable R(t) has to be defined and additional terms have to be included in the right hand side of equation (12) and/or equation (5). The formulation of the discharges of the debt and the availability of new investments cause slightly more complications, but the importance of this part is for the moment considered to be too small to receive more attention. As soon as monetary aspects are introduced, this has to be elaborated of course.

2.6. OBJECTIVE FUNCTIONS

The preceding subsections described the structural relations within our core model. The objectives which are pursued are still to be described. They are of course dependent on the primary purposes for which the model will be used. We will assume the following priorities and their related objective functions:

a. the aim of economic growth within the given parameter values leads to the goal function:

$$Max\{ \sum_{i,t,\tau} FREV(i,t, \ ) - \sum_{i,t\tau} FSUR(i,t) - \sum_{i,t} FINV(i,t)\} \qquad (16.1)$$

which implies the maximization of retained profits payable to share holders.

b. the aim to get a maximum labour income leads to:

$$Max \sum_{t} FINC(t) \qquad (16.2)$$

c. the aim of a maximum employment leads to:

$$Max \sum_{s,i,t,\tau} QLAB(s,i,t, \ ) \qquad (16.3)$$

or, weighted with wage rates:

$$\text{Max} \sum_{s,i,t,\tau} W(s,t)*QLAB(s,i,t,_\tau) \qquad (16.4)$$

Finally, we may also assume a kind of decision game between actors in the model, for example, producers and consumers, each with their specific own goal function. Owing to the fact that the model is strictly linear the game-theoretic approach can easily be applied to the model. Here we restrict ourselves to a simple goal function or a linear combination of two simple alternative goal functions.

3. INTRODUCTION OF POLLUTION AND ENERGY

The preceding section has described a programming reformulation of the main features of an economic model. Recently, these features have been adjusted to new problems: pollution and energy.

3.1. POLLUTION

For the pollution problem we can define one (or more, if detailed description of this aspect is needed) row of coefficients in the primary input sectors of the Input-Output-relations:

$$QSO2(i,t) = S(i,t)* \sum_{t} QP(i,t,\tau) \qquad (17)$$

where QSO2(i,t) denotes the amount of pollution caused by production. The coefficients S(i,t) may be assumed to obey the trend

$$S(i,t) = SO(i)\ EXP(C_i.t), \qquad (18)$$

SO(i) and C_i being parameter values.
The same holds for pollution by consumers:

$$QSO2(c,t) = S(c,t) \sum_{i \varepsilon C} FCEM(i,t) \qquad (17')$$

with a time trend (18') for S(c,t) analogous to (18).

This amount of pollution can be reduced by using special facilities: purifying investments. They provide the capacity for purifying activities reducing the primary level of pollution:

$$Q02(i,t,_\tau) \leq \frac{FPI(i,\tau)}{PINV(\tau)*\lambda(,\tau)} \quad , \ t - \tau < T_S) \qquad (19)$$

where the index i denotes sectors of industry and the consumption sector, Q02(i,t,τ) the amount of pollution reduced by purifying activities in sector i in period t using purifying investments of period which have a technical lifetime of T_S.

The investments, however, affect the funds for productive investments, so that the left hand side of restriction (8) has to be adjusted with an analogous summation of FPI(i,τ), where the index i is referring to sectors of production.

The restrictions put by society on pollution can be described by the right hand side of (20):

$$\sum_i QSo2(i,t) - \sum_\tau Q02(i,t,\tau) \leq S02(t) \qquad (20)$$

3.2. ENERGY

We can include energy consumption either in the way as we formulated the use of labour - via year-of-investment dependent coefficients - or as we did with pollution - via year-of-production dependent coefficients. We have chosen here for the first approach (cf. equation (3) and (4)):

$$QEN(i,t,\tau) = \gamma(i,t)*QP(i,t,\tau) \quad , \qquad (21)$$
$$i \in (\text{production + consumption})$$

where

$$\gamma(i,\tau) = E_i \exp F_i.\tau \qquad (22)$$

When more attention has to be paid to the energy problem, equation (21)-(22) can be subdivided into more categories of energy.

Now the problem arises whether we have to continue with writing down corrective measures as we did with pollution (cf. equation (19)) or whether we have to assume that the use of energy does not leave possibilities for energy saving activities. Assuming that energy saving activities are possible, the best way to formulate the effect of investments in that case should be via an influence on equation (22), but that again violates the linearity of the model. Finally, we have decided to use a formulation analogous to equation (19):

$$QES(i,t,\tau) \leq \frac{FEI(i,\tau)}{PINV(\tau) * \delta(i,\tau)} \qquad (23)$$

where the same consequences as mentioned for FPI hold for equations (8) and (15) with reference to FEI. Finally we formulate the impacts of purifying activities on energy consumption and of energy saving on pollution:

$$QEP(i,t) = EP(i) * \sum_\tau QES(i,t,\tau) \qquad (24)$$

$$QPE(i,t) = PE(i) * \sum_\tau Q02(i,t,\tau) \qquad (25)$$

where in both equations the symbol i refers to sectors of production and the consumption sector.

Consequently, the left hand side of equation (21) has to be adjusted within the brackets by either + QEP(i,t) or - QEP(i,t), the sign depending on the influence assumed: saving energy can have polluting effects of purifying effects. The influences can be made explicit, when the model is extended with one or more energy production equations. This is one of the future research directions.

Finally, society is assumed to set a restriction on the total use of energy

$$\sum_i \sum_\tau QEN(i,t,\tau) + QPE(i,t) - \sum_\tau QES(i,t,\tau) \leq EN(t)$$

4. BALANCE OF PAYMENTS

As mentioned before – see our remarks related to equation (13) – we cannot formulate a programming model which guarantees an equilibrium with the outside world. Nevertheless, it is worthwhile to define some relationships reflecting the balance of trade, viz. imports and exports.

4.1. Imports

Primary imports can be easily defined for sectors of industry:

$$FIM(i,t) = PI(i,t) * \beta_{i,k} * \sum_{\tau} QP(i,t,\tau) \qquad (27)$$

where PI(i,t) are import prices related to the import structure of sector i and $\beta_{i,k}$ technical coefficients. The same holds for autonomous imports of consumption:

$$FIM(i,t) = \varepsilon(i,t) * FDEM(i,t) \quad . \qquad (28)$$

This relationship reduces the demand for home-produced goods with the same quantity at the right hand side of equation (13).

Besides these technological and behavioral effects, we have mentioned already the induced imports FIMP(i,t) in equation (13).

One aspect influencing imports and exports has to be paid attention to, viz. the interpretation of FSUR (cf. equation (7)).

So far we have not made any difference whether we consider FSUR as payment in money to the consumption sector or as a real payment by reducing the prices. In the first interpretation, the effect is purely national, but in the second interpretation the effect influences autonomous imports of consumption (equation (28)) as well as the demand for exports.

So we define parameters (i,t) within the range 0 to 1 denoting the fraction of FSUR(i,t) paid in real terms.

The effect on autonomous imports of consumption goods leads to a change of (28) into:

$$FIM(i,t) = \varepsilon(i,t)*FDEM(i,t) - \\ - \bar{\varepsilon}(i,t) * \Psi(i,t)*FSUR(i,t) \qquad (28')$$

4.2. Exports

Demand for exports is mainly autonomous, which – for this model focusing on a national economy – is the same as exogenous. The price elasticity for exports may be supposed to exceed 1 (in absolute value); so we get

$$FEX(i,t) = FEXO(i,t) + PEL(i)*\Psi(i,t)*FSUR(i,t) \qquad (29)$$

where PEL(i) denotes the final effect of demand for exports on the relative importance of a price reduction. The amounts of FEX(i,t) have to be added to the right hand side of (13).

4.3. Capital Accounts

The part of (dis-)savings which are not used (provided) within the national economy are, by definition, the balance of the capital accounts, but for the moment they are beyond the scope of this research project.

5. PUBLIC SECTOR

So far, we have not paid any attention to the public sector (central and local authorities). One of the aims of this model, however, is to get an idea about the possibilities for governmental policy with respect to the interdependencies between employment, pollution and energy consumption.

So we will introduce income and expenditures for the public sector, which of course will require adjustments for many preceding equations. In order to avoid a reformulation of all these equations in this section, we only mention here the consequence.

5.1. Public Income

a. Taxes on wages:

$$FTAXW(t) = TAXW(t)* \sum_{s,i,\tau} W(s,t) + QLAB(s,i,t,\tau)$$

These amounts have to be subtracted from the right hand side of equation (12).

b. Taxes on profits:

$$FTAXP(t) = TAXP(t)+\sum_{i}[\sum_{\tau} FREV(i,t,\tau)-FSUR(i,t)]$$

These taxation effects have to be substituted properly into equations (5), (6), (7), (8) and (9).

c. Taxes on pollution:

$$FTAXS(t) = TAXS(t)* \sum_{i} QSO2(i,t)$$

These taxes lead to analogous adjustments as FTAXP.

d. Taxes on energy:

$$FTAXE(t) = TAXE(t) + \sum_{i,\tau} QEN(i,t,\tau)$$

They also lead to the same corrections as FTAXP.

e. Other taxes:

$$FTAXR(t) = TAXR(t)* \sum_{i} P(i,t)*QP(i,t)$$

which constitute a burden on the consumption sector, and hence have to be subtracted from the right hand side of equation (12).

5.2. Public expenditures

a. Public employment

The public sector also leads to employment, partly exogenous, partly endogenous:

$$FEMPL(s,t) = W(s,t) * QEMPL(s,t) \geq EMPL(s,t)$$

These amounts are family income (equation (12)).

b. Secondary income
Where the social security sector is lacking in this model, the secondary incomes are assumed to be paid by the government:

$$FSEC(t) = FSEC0(t) +$$

$$+ \sum_{s} 0.8*W(s,t)* \{LAB(s,t) - \sum_{i,\tau} QLAB(s,i,t,\tau\}$$

These amounts are also family income (equation (12)).

c. Public expenditures on goods and services
The public sector asks for goods and services, both endogenous and exogenous:

$$FPDEM(i,t) \geq PDEM(i,t)$$

These amounts are final demand (equation (11)).

d. Investment subsidies
The government can try to stimulate the various types of investments:

$$FGRI(i,t) = GI(i,t) * FINV(i,t) \qquad i \in \text{industry}$$

$$FGRP(i,t) = GP(i,t) * FPI(i,t) \qquad i \in (\text{industry} + \text{consumption})$$

$$FGRE(i,t) = GE(i,t) * FEI(i,t) \qquad i \in (\text{industry} + \text{consumption})$$

These subsidies reduce the effective burden of investment outlays for the recipients in equations (8) and (12).

5.3. Budget
The public sector has to satisfy a budget restriction

$$EXPEND \leq (1.06) * PINC \quad ,$$

where is an exogenous parameter reflecting the maximum shortage on the public budget.

6. CONCLUSION

After the extension of the model in sections 3-5, we have a wider range of options for defining the objective functions presented in section 2.6. We can include in the objective functions additional terms reflecting our preference or dispreference for certain relevant factors such as pollution, energy, balance of payments, and public budget. As a whole, we may conclude that the above mentioned model provides a fairly integrated and broad picture of the functioning of an economy.

9 Economic Structure and the Environment: Production, Pollution and Energy Consumption in the Netherlands, 1973/1985

L. HORDIJK, H. M. A. JANSEN, A. A. OLSTHOORN, J. B. OPSCHOOR, H. F. M. REIJNDERS, J. H. A. STAPEL AND J. B. VOS

I. INTRODUCTION, BACKGROUND, AIMS AND METHOD OF THE STUDY

A. INTRODUCTION

This paper is an abridged version of a report (Hordijk et al, 1979, a) written by a team consisting of economists and engineers at the Institute for Environmental Studies of the Free University, Amsterdam.

The research was sponsored by the Ministry of Public Health and Environmental Affairs.

The study concerns chemical pollution of air and water, production of solid waste, and energy consumption as related to production activities of goods and services. Neither pollution generated by consumption activities nor other types of environmental deterioration are being considered.

This report is the follow up to "Environmental Pollution and the Economic Structure in the Netherlands" (Jansen et al., 1978), to which some improvements are made. A difference lies in the model used: the input-output model now contains an energy-part in which production and use of various types of energy have been treated in detail. The economic part of the model is more disaggregated now. Also, the emission data has been improved on the basis of new information. Moreover a new trendscenario for 1985 is constructed. Finally, technological innovation in the period 73/85, expressed in the form of changes in future emission coefficients, has been treated more extensively.

B. BACKGROUND

In the 20th century production activities in the Netherlands showed an almost explosive growth. From the beginning of the century population almost tripled and total real GNP grew with a factor 10. Most of this growth was brought about after the Second World War: in the period 1900 - 1950 GNP tripled and from 1950 until 1980 it tripled again. In the postwar period environmental pollution increased correspondingly, although concern about the environment was growing as well, particularly in the sixties and early seventies.

This concern lead to an increasing call for the institution of environmental policies. In the Netherlands, then the Ministry for

Public Health and Environmental Affairs was separated from the Ministry for Social Affairs.

In cooperation with other Ministries, directly involved (e.g., Transport and Public Works) or indirectly involved (e.g., Economic Affairs), this Ministry devised certain policies which appeared to have had some success. From about 1970 onwards, SO_2-emissions decreased (this was, however, mainly due to a fuel switch from imported oil to domestic natural gas) and emissions of biodegradable matter in water diminished considerably. In 1972 the Ministry announced a clean-up period for air and water in order to reach acceptable levels of pollution in 1985. A legislative framework has been constructed and is being extended in order to provide government with instruments to control pollution of air and water, solid waste, chemical waste and noise.

However, there are good grounds for a pessimistic view concerning long term prospects. The downward trend of SO_2-emissions in the past is not representative for total pollution. It is expected that in the future depletion of domestic natural gas resources will lead to an increasing consumption of coal and oil and thus to an increase of SO_2-emissions.

A steady growth (even at a rate lower than in the sixties) will soon lead to a situation where existing standards with respect to traditional indicators of pollution like SO_2 and NO_x are exceeded (not to mention "new" pollutants such as for example nuclear waste). An additional draw-back has become that political economic issues (recession, unemployment) tend to overshadow environmental problems.

This paper deals in particular with socalled structural abatement of pollution, which must be distinguished from technological abatement (i.e., changes and improvements of production processes), legislative measures, spatial planning, etc. Structural abatement (see James et al, 1978) is aimed at attaining a more acceptable situation through changes in the pattern of economic activities. In the Netherlands this has been called "selective growth", which means that the growth rates of the various economic activities are differentiated in such a way that a variety of targets are better approximated, such as with respect to environment, natural resources, income, employment, quality of labour and third world problems. In 1976 the Ministry for Economic Affairs published a Memorandum on Selective Growth (Economische Structuurnota, 1976). In this Memorandum it was stated that the so-called oriented market-economy ("economie du marche oriente") constitutes the existing economic order in the Netherlands. Within this economic order a policy of directly influencing production levels was rejected. The minister expressed preference for global instruments, and while taking into account targets which he calls "facets" (environment, international division of labour, quality of labour, third world problems) top priority has been given to income, employment and repelling inflation.

One of the problems the Memorandum stipulates with respect to the target "environment" is lack of information. It has been a purpose of this study to fill this gap to some extent.

C. AIMS AND METHOD OF THE STUDY

The first aim of this study has been to analyze the relationships between the production pattern, pollution and energy consumption by reviewing the situation in 1973, the most recent year for which a detailed economic input-output table has become available. In order to attain as much detail as possible the analysis has been applied at the highest possible level of disaggregation, namely:

- 60 economic sectors (usually 23 sectors are distinguished). This number was required in order to get a classification, more or less homogeneous with respect to both economic and environmental aspects;

- 17 types of energy. These were divided according to use (combustion/non-combustion) into 32 energy sectors;

- 55 pollutants as emitted by the economic sectors.

For each sector of this classification (for the key see Appendix A) data has been collected of emission/energy consumption. To make this extensive information more concise for the reader, air pollution and water pollution by heavy metals were aggregated into the "Air Pollution Equivalent" (APE) respectively the "Heavy Metals Equivalent" (HME), by weighing the various chemicals with a factor reflecting as accurately as possible their detrimental effects.

A second goal has been to attain insight into pollution and energy consumption connected to the final products of the productive sectors. This includes not only pollution and energy consumption generated in the last phase of production by the sector delivering the final product, but also the pollution generated in previous production phases by sectors delivering intermediate products. Input-output analysis has been used to compute the cumulated coefficients, indicating total pollution and energy consumption per unit of final product. With this information, a rough classification of the economic activities can be drawn up, separating economic sectors according to pollution intensity.

In this connection, the impact of foreign trade on the natural environment in the Netherlands and abroad has been analyzed.

Finally, a scenario-analysis for the year 1985 has been undetaken, investigating the policy margins of selective growth with respect to the facet "environment". Using extrapolation of recent trends as well as predictions and taking account of policy intentions, a trendscenario can be constructed for the production level of each sector. For emissions and energy consumption this required estimates of the 1985-coefficients. Contrasting scenario's were also constructed by the use of linear programming, with constraints on environmental and on economic targets. These contrasting scenario's give some idea of the effectiveness which might be expected if a selective growth policy were directed to a greater extent to the facet of envioronment without the neglect of economic aspects like income and employment.

II. INPUT-OUTPUT ANALYSES AND SOME ENVIRONMENTAL APPLICATIONS

A. INTRODUCTION

As a first approximation of the relationship production/ pollution one could state that an increase in production with x% leads to an increase of pollution with the same percentage. However, in the context of selective growth modelling this approach will not do, as productive sectors differ with respect to environmental impact. Moreover, production takes place in sectors which are mutually connected. In consequence, expansion or reduction of one sector will affect production and pollution in all other sectors, which, in turn, will cause subsequent cross-effects.

Input-Output Analysis (IOA), a research method developed by Leontief (1951) is concentrated on those mutual dependencies between economic sectors. This method of analysis, although originally developed for purely economic problems, is very well applicable to the analysis of polluting effects of production, cf. inter alia Den Hartog and Houweling (1974), Leontief (1970) and Victor (1972).

IOA investigates the effects of a change in final demand (i.e., goods and services bought by consumers, government and export, or stock changes) on production levels. For both final demand and production a distinction is made between the various sectors. The connection between the various economic sectors proceeds through their intermediate deliveries. For further information we refer to Dorfman, Samuelson and Solow (1958), James et al. (1978) and Leontief (1951).

B. EXTENSION OF IOA WITH ENERGY SECTORS

In IOA the product flows are usually measured in monetary units. This raises problems where price differences exist for various buyers: monetary units do no longer reflect physical units. For energy there are considerable price differences for different buying sectors. To overcome this problem, IOA has been extended with special energy sectors (see also Metra, 1978). These sectors are supplied in monetary units by the sectors "petroleum and coal products", "electric utilities", "natural gas and petroleum" and "coal mining" and they in turn supply in physical units (joules) to the energy consuming sectors.

The introduction of these energy sectors also creates the possibility to pay attention to changes in the mix of energy consumption by sector, by changing the appropriate input-output coefficients. In this way a future switch of natural gas to coal and oil, for example, can be readily accounted for in the model.

C. PRIMARY AND CUMULATED COEFFICIENTS

When estimates of emission per sector are made (see chapter 3), emissions can be related to the levels of production. Supposing that pollution and energy consumption per unit of production is constant per sector, we have an indicator of the polluting impact of a sector by taking the quotient of emission and production value, i.e., the primary pollution or energy coefficient.

However, the ultimate goal of production is meeting final demand. To determine the pollution content of a final product, it is not sufficient to take the pollution generated by the final product sector. This sector has made use of goods and services delivered by other sectors and in those sectors pollution may have been generated as well. To determine the total pollution or energy content of a final product not only pollution generated or energy used in the last phase of production but also that of previous phases must be taken into account. Following the Dutch Central Planning Bureau (Den Hartog & Houweling (1974) we call the total pollution/energy consumption per unit of final product the cumulated coefficients.

D. LEVEL OF AGGREGATION

Traditionally, the decision about the level of aggregation of productive sectors is based on economic considerations. Activities are combined in such a way that the resulting sectors are, as much as possible, homogeneous from an economic point of view. However, this does not guarantee homogeneity from an environmental point of view. For this reason a number of sectors has been redivided in order to attain more environmental homogeity.

E. TWO APPLICATIONS: SCENARIO ANALYSIS AND IMPACT OF FOREIGN TRADE

Scenarios depict possible future realizations of, for instance, economic and environmental developments. The scenario to be expected in the case of unchanged present policies can be compared with possible implications of policy changes.

IOA has been employed to obtain consistency in the scenarios, with respect to the mutual interdependencies between sectors. A problem thereby is, that future technical relationships between sectors are not known (in this case, for the year 1985). Instead, 1973 data had to be used. It was possible however, to take into account some changes, notably with regard to fuel switches.

With the aid of IOA coupled with the associated pollution coefficients, the amounts of pollution for export products have been calculated as generated in the Netherlands. Likewise, it is possible to calculate pollution abroad, generated by the importation of goods and services into the Netherlands. The balance of these amounts yields the "balance of pollution" as a counterpart of the balance of trade (Vos, 1978).

III. POLLUTION AND ENERGY CONSUMPTION PER SECTOR, 1973

A. ENERGY CONSUMPTION PER SECTOR

The data for energy consumption per sector has been provided in a non-published energy balance for 1973, furnished by the Dutch Central Bureau of Statistics. These figures were disaggregated and corrected on the basis of available newer information, while for some fuels special estimates had to be obtained.

The figures are presented in an aggregated form in Table 1. For more detailed information we refer to Hordijk et al. (1979).

B. AIR POLLUTION PER SECTOR

B.1. Introduction

The Emission Registration of TNO (a technological research institute, head office Delft) commissioned by the Ministry of

TABLE 1.

ENERGY CONSUMPTION (PJ) BY SECTORS, 1973

Sectors / Types of energy													
Coal + coalproducts	0.2	-	-	-	2.8	2.3	0.4	1.3	27	0.23	25	-	5.2
Coal + coalproducts, feedstock	-	-	-	-	23	-	4.7	-	45	0.4	-	-	-
Natural gas	58	53	9.2	23	0.7	46	88	37	26	22	344	13.4	93
Natural gas, feedstock	-	-	-	-	-	61	1.2	-	-	-	-	-	-
Oilproducts	48	15.1	3.5	1.2	174	2.2	56	6.5	17.1	8.2	57	20.1	48
Oilproducts, feedstock	-	-	-	-	-	-	214	-	-	5.7	-	-	-
Electricity (public utilities)	0,28	7.0	2.6	2.6	3.2	4.3	18	3.6	18.5	10.9	-	6.8	31
Petrol and diesel (transport)	-	14	1.3	1.0	-	0.2	3.4	0.9	1.0	6.6	-	3.6	180
Number of the sector (see Appendix A)	3	9 - 18	19 - 22	26/ 27	29	30	31/ 32	34	35	36/ 40	42	6/8, 23/25 28,33 41	1,2,4, 43- 60

Public Health and Environmental Affairs is an important data source for emissions of air pollutants. In this registry several hundred pollutants have been distinguished. Firms cooperate on a voluntary basis to furnish data on emissions. Unfortunately, the data were not yet fully completed when this report was written. For our purpose we have distinguished 14 types of air pollutants:

carbonmonoxide, sulphurdioxide, nitrogenoxides, chlorinated and/or fluorinated hydrocarbons, aromatic hydrocarbons, hydrocarbons containing oxygen, hydrocarbons containing sulphur or nitrogen, unsaturated hydrocarbons, saturatred hydrocarbons, fluorine, mercury, lead, other anorganic compounds, anorganic particulate matter and organic particulate matter.

B.2. Combustion Emisions (transport excluded)

In order to calculate sulphurdioxide emissions, energy consumption has been multiplied by emission factors. Emissions of nitrogenoxides depend on fuel, furnace construction, the type of generator, etc. and also on the firm involved. On the basis of the Emission Registration data various emission factors of NO_x can be derived (Werkgroep NOP_x - vorming vuurhaarden 1978). Other emission factors have been constructed by the the Central Bureau of Statistics(Tinbergen and Schake, 1977). Estimates of emissions are given in an aggregate form in Table 2.

B.3. Process emissions

Process emissions are all non-combustion emissions. Roughly speaking they can be distinguished into three categories:

(i) waste products of chemical processes, for which recycling or use as feedstock is too expensive (for instance SO_2 and NO_x in the production of sulphuric or nitric acid);

(ii) leakage, spilling, etc. during transport of processing;

(iii) solvents, for instance as in paint, evaporating during use of the product.

Information on process emissions is scarce. The reliability of our estimates is therefore variable. Figures, as they have been compiled, are given in an aggregated form in Table 3.

B.4. Transport emissions

For transport emissions it is not sufficient to multiply energy consumption by emission factors because these factors differ for the various means of transport.

In the input-output table most transportational services are delivered by the sector "other transportation" to "wholesale trade"; "wholesale trade" delivers transportation services together with other services to the other sectors. This leads to problems, as it is not possible, with available data, to make a distinction between transportation services and other services (the monetary units do not reflect very well the physical transportation deliveries). That is why transportation emissions have been calculated exogenously and assigned directly to the receiving sectors. Results can be found in an aggregated form in Table 4.

TABLE 2.

COMBUSTION EMISSIONS (tons, 1973), TRANSPORT EXCLUDED

Sector (ton/emission)	CO	SO_2	NO_x	Hydrocarbons containing oxygen	Saturated Hydrocarbons	Particulates
Horticulture	84	57000	9500	250	500	2600
Food and beverages	27	15000	4800	75	140	750
Textile	3	4000	1100	18	35	200
Paper	1	1500	3400	6	13	80
Petroleum and coal products	10	192000	20200	600	1500	5700
Fertilizers	2	2600	8500	11	20	170
Other chemical products	450	42000	15000	440	800	1500
Building materials	100	7200	3300	430	60	570
Primary metals	72	23000	10000	1000	770	3000
Other metals	100	7800	4200	240	80	500
Electricity	500	65000	67000	700	900	1900
Other manufacturing	1500	26000	5200	260	9200	1300
Other	1300	16000	11000	3300	410	2000
Total	4300	460000	163000	7400	14000	20000

TABLE 3

PROCESS EMISSIONS
(tons, 1973)

emission Sector	CO	SO_2	NO_x	HC(Cl/F)	Aromatic HC	HC(O)	HC (N/S)	unsat. HC	Sat.HC	F	Other anorg. components	Part. anorg.	Part. org.
Agriculture and horticulture 1-9	-	-	-	500	-	-	-	-	-	-	-	-	-
Food and bev. 9-18	-	-	-	-	-	6,000	-	-	6,000	-	-	-	7,600
Textile 19-22	-	-	-	-	600	550	-	-	4,000	-	-	-	-
Paper 26-27	-	-	-	-	1,350	1,350	-	-	-	-	-	-	-
Petr. and coal prod. 29	75	5,600	-	-	1,600	-	-	-	-	-	-	-	-
Fertilizers 30	-	15,500	19,900	-	-	-	-	-	38,000	-	130	2,450	-
Other chem. basic products 31	24,000	19,000	4,000	6,500	2,500	10,000	500	15,000	4,000	150	5,000	5,200	850
chem.fin. prod. 32	-	-	-	1,000	1,600	4,500	-	150	1,000	-	-	-	400
Build. materials 34	12,000	450	-	-	-	-	-	-	-	1,300	200	8,600	-
Primary metals 35	165,000	9,300	3,350	-	350	-	-	-	-	950	190	13,900	-
Other metals 36-40	2,700	-	-	4,265	9,600	1,520	-	-	28,250	-	-	14,000	-
Other manufacturing 41,42,33,28, 6,7,8,23,24, 25	-	-	-	2,600	750	7,350	-	-	3,500	-	-	1,400	3,000
Sea & air nav. 50	-	-	-	100	450	400	100	-	15,000	-	-	-	-
Wholesale, ret.trade 47-48	-	-	-	-	-	-	-	-	21,000	-	-	-	-
Other	-	2,300	1,100	5,300	4,800	2,600	-	-	34,600	300	4,000	99,500	200
Total	200,000	52,000	28,000	20,000	24,000	34,000	600	15,000	155,000	2,800	15,000	153,000	12,000

TABLE 4.

TRANSPORT EMISSIONS (tons, 1973) PER SECTOR

Sector / emission		CO	NO_x	SO_2	Part.	KW(O)	Lead	Saturated HC
Agriculture and horticulture	1-5	50,000	15,000	2,700	900	100	50	6,700
Food and bev.	9-18	100,000	22,000	3,000	1,100	50	100	11,000
Textile	19-22	8,700	1,800	250	100	1	100	1,000
Paper	26/27	4,000	800	100	50	1	5	500
Pet. and coal prod.	29	9,800	5,000	1,000	300	50	10	1,600
Fertilizers	30	5,900	2,200	450	150	20	10	1,000
Other Chem.bas.prod.	31	9,400	3,900	800	250	34	10	1,500
Chem.fin.prod.	32	5,400	1,000	150	50	5	7	700
Build. materials	34	14,000	12,000	3,000	800	150	10	4,000
Primary metals	35	6,000	1,900	400	100	20	7	900
Other metals	36-40	4,000	6,100	700	300	3	50	4,500
Other manufacturing	41,42,33,28, 6,7,8,23,24,25	4,820	9,400	1,300	500	30	60	5,500
Sea and air nav.	50	-	-	-	-	-		-
Wholesale,ret.trade	47/48	74,000	7,800	600	350	-	100	8,000
Other		290,000	65,000	10,000	3,500	350	350	37,000
Total		670,000	150,000	25,000	8,500	800	800	85,000

C. WATER POLLUTION

Three categories of water pollution are included:

1) Biologically degradable organic waste (BDOW, measured in population equivalents (p.e.)). The main source for this information is found with the Central Bureau of Statistics (CBS, 1978), but the data has had to be desaggregated.

   It has not been possible to determine per sector the amounts of emissions that have been treated in collective water purification installations; however, only 10 to 15% of industrial emissions was treated in 1973 (mainly the sectors "food" and "beverages").

2) Non oxygen-using materials. This category contains heavy metals, electrolytes, mineral oil, etc.

3) Phosphates and nitrogen compounds. In this category industrial emissions are only a minor fraction, as private households and, notably physical imports (rivers Rhine and Maas) prove to be far more important.

Some figures are given in Table 5.

D. SOLID NON-CHEMICAL WASTE

The available information on solid waste has been less, at least during the research period, than that on air and water pollution. There are various ways for treatment and/or disposal of solid waste: controlled tipping in land or in water, incineration, composting, etc. We have only estimated the production of solid waste: it is not clear which part of this amount would in the end add to the pollution of the environment. However, it may be said that in 1973 the treatment capacity for solid waste was far from sufficient.

E. WEIGHING EMISSIONS

Due to the extensiveness of the data on emissions it is difficult to get an overall picture of pollution in general. For this reason indexes were made up for air pollution (Air Pollution Equivalent, APE) and for water pollution by heavy metals (Heavy Metals Equivalent, HME). Weights were chosen to reflect the adverse effects of the pollutants. For air pollution these effects depend on several factors, for instance: quantity and location of the emission, stack height, meteorological conditions, interaction with other pollutants and dose-effect relationships. Because of incompleteness of information on these factors the following simplification has had to be introduced: effect = emission/standard.

As standards, MIC-values (maximum immission concentration) were used. This has yielded the weights for APE, given in Table 6.

TABLE 5.

WATER POLLUTION, 1973

Sector	Sector nr.	BDOW (1000 p.e.)	Mineral oil (ton)	Cadmium (ton)	chroom (ton)	Heavy metals equivalent (see 3.6)
Horticulture and agriculture	1,2,3,4,5	560	-	-	-	2
Food and bev.	9-18	11,000	-	-	-	-
Textile	19-22	540	-	-	3	1
Paper	26,27	2,000	-	-	-	-
Pet. and coal prod.	29	240	1,300	-	-	0.1
Fertilizers	30	400	-	16	17	27
Other chem.bas.prod.	31	3,000	400	0.3	80	75
Chem. fin. prod.	32	200	-	-	100	13
Building materials	34	33	-	-	-	-
Primary metals	35	140	1,200	-	110	180
Other metals	36-40	350	200	10	150	54
Other manufacturing	6,7,8,23,24,25 28,33,41,42	240	-	-	37	5
Sea and air nav.	50	120	-	-	-	-
Wholesale, ret.trade	47,48	500	-	-	-	-
Other*	43,44,45,46,49	3,600	300	-	-	-
Total	51-59	24,000	3,400	26	500	

* emissions of water purification installations included.

TABLE 6.

MULTIPLICATIVE WEIGHTS FOR THE AIR POLLUTION EQUIVALENT.

Pollutant	Weight	Pollutant	Weight
CO	.0018		.02
SO_2	.015	sat.HC	.025
NO_x	.015	fluorine	1
HC(Cl/F)	.4	anorg.compound	.33
aromatic HC	.2	org. part.	.013
HC(O)	.04	anorg. part.	.013
HC(N/S)	.5	lead	1.33

For water pollution by biologically degradable organic waste (BDOW) the population equivalent (p.e.) is a useful and generally accepted aggregate. A rough aggregate for water pollution by heavy metals was made up by using the so-called black and grey list with the weights 1 respectively 0.1. Heavy metals have been black- and grey-listed in Table 7.

TABLE 7.

HEAVY METALS, AGGREGATED IN HME

Black list	Grey list
Hg, Cd	Cr, Zn, Ni, Se, Cu, Pb, V, As

F. APE OF PROCESS, COMBUSTION AND TRANSPORT EMISSIONS

In Table 8 it can be seen how emissions of various air pollutants contribute to APE and how they are distributed between transport, conbustion and process emissions.

TABLE 8.

COMPARISON OF COMBUSTION, PROCESS AND TRANSPORT EMISSIONS OF PRODUCTIVE ACTIVITIES, 1973, 1000 TONS

	APE	SUMMED emissions	CO	SO_2	NO_x	saturated HC	other HC
Combustion	10	670	4.3	460	160	14	28
Process	24	670	180	33	24	150	280
Transport	7.3	940	670	25	150	85	10

When measured in summed unweighed tons (column 2) transport emissions are the most important category, while process and combustion emissions share second place.

Until recently the commonly used indicator of air pollution was sulphurdioxide, which caused an underestimation of process and transport emissions as air pollution sources.

The picture changes strongly by implementation of the proposed weighing procedure. Then, process emissions turn out to be the most important category. In evaluating these figures it should also be kept in mind that contrary to combustion emissions, process emissions mostly take place at low heights, which is important for ambient concentrations. This last factor is not reflected in the weights.

In an memorandum on the Planning of Air Pollution Abatement (Ministry for Public Health and Environmental Affairs, 1977) use has been made of another weighing procedure. Process emissions seemed less important there. This can be explained by the fact that a lower number (five) of pollutants was included. In fact, the more unfamiliar pollutants turn out to be relatively significant. Another reason for the difference is that different weights were employed.

G. EMISSIONS PER SECTOR

In this section we give some of the most important results regarding the emissions of APE, BDOW, HME and energy consumption per sector. Unpublished data, confidentially supplied by the Central Bureau of Statistics has been used in the calculations. Some of the figures have been omitted in Table 9. Cumulative emissions in the table are established by multiplying the cumulated coefficients with final demand.

The corresponding coefficients are given in Table 10; these are emissions per unit of production respectively final demand.

In order to make a classification of "dirty" and "clean sectors, these have been divided into five classes for each of three indices (APE, BDOW, HME). The highest cumulated coefficients were put in class 1, the lowest into class 5. A sector has been called "dirty" if the coefficient for at least two of the three indicators were allocated in class 1, 2 or 3, while it has been called "clean" if at least two coefficients were allocated to class 5 without the third coefficient belonging to class 1, 2 or 3. This procedure has yielded the results of Table 11.

TABLE 9.

ESTIMATED PRIMARY AND CUMULATED EMISSIONS IN 60 SECTORS, NETHERLANDS, 1973

*Sector	Air Pollution (APE)		Water Pollution BDOW(1000 p.e.)		Water Pollution (HME)		Energy Cons. (JP)	
	prim	cum	prim	cum	prim	cum	prim	cum
1. Livestock ext.	110	83	370	92		0.59	2.0	4.8
2. Livestock int.	230	183	130	155	2	0.78	7.0	8.0
3. Horticult.	1130	1175	10	20		0.56	105	107
4. O. agr. cult.	510	424	50	63		1.9	20	22
5. Fishing	-	0.3				0.0	5.5	0.05
6. Coal mining	443	290		2		0.06	30	20
7. Gas	53	105		14		1.4	0.1	11
8. Other m.	25	20		2		0.05	5.0	3.0
9. Meat prod.	120	870	1000	1600		0.04	5.5	41
10. Dairy prod.	260	670	850	1250		3.4	18	42
11. Canning,pres. 19		150	345	330	450		1.2	3.5
12. Grain	330	-	40	-		-	10	-
13. Sugar	100	80	610	230		0.23	10	6.2
14. Flour	60	200	30	450		0.43	5.0	13
15. Sweets	65	-	30	-		-	2.5	-
16. Other f.	430	-	6900	-		-	14	-
17. Bev.	60	160	1400	1200	0.2	0.72	5.0	10
18. Tobacco	50	90	6	27		0.24	2.0	3.8
19. Spinning	32	-	70	-		-	2.0	-
20. Weaving	130	-	7	-		-	5.0	-

Table 9 (continued)

	prim	cum	prim	cum	prim	cum	prim	cum
21. Knitting	22	-	65	-		-	2.0	-
22. Oth.Text.	220	-	400	-	1.25	-	5.0	-
23. Clothing	75	150	25	54		0.17	4.0	7.4
24. Leather	30	60	90	100	3.0	3.0	1.5	2.6
25. Wood	380	310	28	37		0.71	10	11
26. Paper	100	90	2000	950		0.18	23	14
27. Pap.prod.	40	83	10	240		0.29	2.0	6.4
28. Print,publ.	250	220	35	280	1.9	1.3	10	15
29. Petrol.	4800	3600	240	220	0.15	1.8	178	138
30. Fert.	1300	870	400	250	22.7	14.0	112	77
31. Chem.b.pr.	9200	8800	3000	2750	56.1	52	362	386
32. Chem.prod.	560	690	200	290	12.6	12	8.5	19
33. Rubber	1200	800	14	67		0.92	9.5	14
34. Build.mat.	2100	560	33	16		0.35	45	15
35. Pr. met.	2600	2160	140	120	173	125	117	123
36. Met.prod.	700	630	60	75	54.0	35	8.0	20
37. Mach.	540	670	46	78		7.4	6.0	18
38. El.prod.	1000	1280	61	140		7.1	17	40
39. Auto	45	210	16	44		3.1	2.5	9.7
40. Other transp.		624	940	165	200	11.0	4.5	22
41. Other manuf.		24	70	35	37	1.0	2.0	4.8
42. Electr.	2200	700	15	11		0.17	425	130
43. Gas	40	42	5	4		0.4	11	3.5
44. Water	-	22	5	7		0.13	0.2	2.6
45. Build.	2900	5000	230	430		19.4	14	98
46. Wholes.tr.	1500	1500	150	250		1.7	18	47
47. Ret. tr.	1000	1400	220	420		1.8	32	56
48. Hotels,rest., bars		190	330	270	650	0.55	10	20

Table 9 (continued)

	prim	cum	prim	cum	prim	cum	prim	cum
49. Rep.	460	360	100	72		1.6	6.5	9.4
50. Sea/air	440	775	120	170		0.72	8.5	94
51. Other trans.		460	500	100	100	1.1	55	42
52. Commun.	25	50	35	25		0.33	5.0	4.7
53. Bank	120	200	70	100		0.65	9.2	16
54. Real est.	-	270	10	36		1.2	-	5.7
55. Bus services		80	75	70	52	0.15	8.0	6.6
56. Soc.Services		70	95	225	180	0.31	7.4	7.8
57. Medical	175	400	225	350		1.1	22	35
58. Culture, recr.		23	45	18	28	0.14	5.0	6.0
59. Other serv.	2200	1920	3500	3300		0.22	6.0	9.1
60. Other	-	17	-	8		0.18	-	1.3
Total	41600	41600	24200	24200	327	327	1910	1910

* For key see appendix.

TABLE 10.

PRIMARY AND CUMULATED EMISSION COEFFICIENTS, NETHERLANDS, 1973

Sector	Air Pollution ($APE/10^6Dfl$)		Water Pollution BDOW ($1000p.e./10^6Dfl$)		Water Pollution ($HME/10^6Dfl$)		Energy Cons. ($PJ/10^6Dfl$)	
	prim	cum	prim	cum	prim	cum	prim	cum
1. Livestock ext.	16.8	85.0	56.8	95		605	0.30	4.9
2. Livestock int.	38.4	131	21.1	111	325	562	1.15	5.7
3. Horticult.	596	651	5.3	11		310	56.2	59.5
4. O.agr.cult.	161	257	15.7	38		1161	6.2	13.1
5. Fishing		94.0		32		294	12.8	16.6
6. Coal min.	1684	2130		14.7		440	114	147
7. Gas	14.3	66		8.7		870	0.03	6.9
8. Other m.	60.4	131		14		352	12.3	19.4
9. Meat pr.		130		241		590		6.1
10. Marg.,oil, other food		147		272		747		9.0
11. Canning.pres.		182		242		635		10.0
12. Grain								
13. Sugar		249		690		684		18.7
14. Flour		105		236		220		6.6
15. Sweets								
16. Other f.								
17. Bev.		84		660		380		5.5
18. Tobacco		42		12.8		115		1.8
19. Spinning								
20. Weaving								

Table 10 (continued)

	prim	cum	prim	cum	prim	cum	prim	cum
21. Knitting								
22. Other tex.								
23. Clothing	31.8	65.9	10.6	24		78	1.7	3.3
24. Leather	45.2	100	134	174	4477	5136	2.1	4.5
25. Wood	110	168	8.0	20.5		389	2.9	6.2
26. Paper		154		1607		301		24.0
27. Pap.prod.		123		359		437		9.6
28. Print,publ.	43.8	101	6.1	127	329	602	1.8	6.6
29. Petrol.	502	510	24.9	31	16	255	18.4	19.5
30. Fertiliz.	1020	1182	313	340	17748	19017	87.8	105.5
31. Chem.b.pr.		1240		401		7540		56.4
32. Chem.pr.		248		105		4324		6.8
33. Rubber		636		53.2		727		11.3
34. B.mat.	613	713	9.6	21.0		443	13.2	19.6
35. Pr.met.	281	626	14.9	35.9	18495	36351	12.5	35.8
36. Met.pr.		197		23.6		10960		6.2
37. Mach.		132		15.4		1470		3.5
38. El.prod.	102	147	6.1	15.8		827	1.7	4.6
39. Aut.man.		106		22.6		1610		5.0
40. Oth.tr.		210		45.6		2452		5.0
41. Oth.man.	18.2	67.7	26.4	35.9		955	1.5	4.6
42. Electr.		433		6.8		105		80.3
43. Gas		30		2.9		280		2.5
44. Water		43.3		13.6		247		5.0
45. Build.	117	243	9.2	21.0		941	0.5	4.7
46. Wh.trade		118		19.3		133		3.7
47. Ret.trade		128		37.4		163		5.0
48. Hot.,rest., bars	66.0	132	92.1	257		219	3.6	7.9
49. Rep.	151	200	32.8	40.3		871	2.1	5.3

Table 10 (continued)

	prim	cum	prim	cum	prim	cum	prim	cum
50. Sea/air	94.1	177	25.7	38.2		165	17.8	21.5
51. Oth.tr.	48.8	107	10.5	21.2		245	5.8	9.1
52. Commun.	6.8	30.3	9.6	15.7		205	1.4	2.9
53. Bank	21.3	68.7	12.8	34.7		229	1.7	5.5
54. Real es.		38.4	1.4	5.1		174		0.8
55. B.ser.	16.9	19.3	14.9	20.5		60	1.7	1.6
56. Soc.ser.	35.7	68.6	113	126		219	3.7	5.5
57. Medical	19.6	45.8	24.9	39.9		132	2.4	4.0
58. Cult.rec.	15.6	38.8	12.5	24.0		122	3.3	5.0
59. Oth.ser.	523	543	928	932		63	1.6	2.6
60. Other	-	85.6	-	39.2	-	886	-	6.7

TABLE 11.

CLASSIFICATION OF SECTORS

"Clean"	Sector nr.	"Dirty"	Sector nr.
Water	44	Coal mining	6
Other trans.	51	Fertilizers	30
Real estate	52	Chem.b.prod.	31
Commun.	54	Build.mat.	34
Wholes.trade	46	Prim.metals	35
Med.services	57	Paper	26
Tobacco	18	Met.prod.	36
Cult.,recr.	58	Marg.,oil, other food	16
Gas	43	Sugar	13
Fishing	5	Chem.prod.	32
Bank	53	Other transp.	40
Clothing	23	Canning,preserving	11
		Other textile	22
		Leather, shoes	24
		Elect.prod.	38
		Hotels, rest.,bars	48
		Other serv.	59

IV. POLLUTION AND THE ECONOMIC STRUCTURE, 1973

A. INTRODUCTION

In this chapter we explore the relationships that can be established between pollution characteristics and the economic structure. First we relate adverse pollution effects and energy consumption to favourable economic effects on income and employment. Then, the analysis questions, to what extent structural interrelations reflected by intermediate deliveries are connected to pollution. Finally, we consider the environmental effects of international trade.

B. POLLUTION, ENERGY CONSUMPTION AND ECONOMIC INDICATORS

When judging the impacts of the productive activities of the economic sectors, one should consider favourable as well as adverse effects. Of the favourable effects we have selected to concentrate on income and employment. Then, the ratios of cumulated coefficients of pollution or energy consumption and income or employment have been calculated. It turned out that the indicators income and employment yielded both roughly the same results.

In table 12 those sectors have been mentioned for which the ratio of cumulated coefficients of pollution or energy consumption and employment was more than twice or less than half the overall national ratio.

In Table 13, we present the corresponding figures for a selected number of sectors, to give an impression of the variation. Polluting and energy consuming sectors turn out to be often labour-extensive.

TABLE 12.

SECTORS (for key see appendix) WITH POLLUTION/ENERGY CONSUMPTION (cumulated) MORE THAN TWICE OR LESS THAN HALF NATIONAL OVERALL POLLUTION/ENERGY CONSUMPTION PER MAN YEAR (cumulated)

	less than half national average	more than half national average
air pollution, combustion (APE)	2,4,5,7,9-22 25,27,28,32,33 36,38,39,40,51	3,6,8,26,29,30, 31,34,35,42,43 50,60
air pollution, process (APE)	1,2,3,5,8,9,10, 11,14,15,17,18,19, 21,23,24,25,26,28, 42,44,46,47,48, 51-58	29,30,31,32,33, 34,35,50,59
water pollution (BDOW)	1-6, 8-23, 25,26,27,28,34,41 ,42,44-59	24,30,31,32,35
water pollution (HME)	3,4,5,6,7,8,23, 25,34,36,37,38,39, 41,42,43,44,45,46, 47,49,51-58	11,12,13,16,17, 22,26,27,29,30, 31,59
energy consumption (combustion)	1,23,24,37,45-49, 41,52-59	3,6,8,26,29,30, 31,34,35,42,43,50, 60

TABLE 13.

INDICES OF (cumulated) POLLUTION OR ENERGY CONSUMPTION PER (cumulated) EMPLOYMENT FOR SOME SECTORS (1973); NATIONAL AVERAGE = 100

Sector	Energy Consumption (combustion)	Air Pollution (APE)	Water Pollution (BDOW)	Water Pollution (HME)
Horticulture	1090	25	13	26
Other mining	374	36	17	32
Dairy products	73	30	164	32
Margarine, oil, other foods	120	83	2021	30
Lumber and wood products	71	48	16	21
Paper	412	46	1838	24
Petroleum and coal prod.	2391	1203	257	149
Other chemical basic products	684	1705	656	876
Build. materials	259	443	18	27
Fab.metal prod.	60	112	18	597
Electr. utilities	3080	21	17	18
Communications	33	10	13	11
Social services	14	3	21	3

C. PRIMARY COEFFICIENTS VERSUS CUMULATED COEFFICIENTS

In Table 10, it was shown that primary and cumulated coefficients differ considerably. This reflects the fact that for certain final products most of the generated pollution did not take place in the last phase of production but in previous phases. To analyse this effect we have used the measure n_k, so that:

$$n_k = \sum_{j=1}^{n} (c_{kj} - p_{kj}) \cdot Y_j / E_k$$

where k indicates the type of pollutant
$j=1,..,n$ indicates the economic sectors
c_{kj} and p_{kj} indicate cumulated and primary coefficients
Y_j final demand
E_k total emission of pollutant k

The values which n_k can take are between 0 (all pollution takes place in the last phase of production) and 1 (all pollution is generated in previous phases).

As the fraction of final demand in total production turned out to be about 35%, it was not surprising that most values of n_k were found to be between .3 and .4. High values were found for combustion emissions (APE), energy consumption, water pollution (HME), process emissions of SO_2, CO, organic particulate matter and organic solid waste. Especially for these pollutants the structural interdependencies of the economic sectors as specified in IOA prove to be important.

D. BALANCE OF POLLUTION AND OF ENERGY CONSUMPTION

In the Netherlands foreign trade plays an important role in the economic process. From the balance of payment we have derived a balance of pollution. At the debit side of this balance we put emissions connected with the export of goods and services (transit excluded). At the credit side appears pollution generated abroad connected with Dutch final and intermediate imports (transit excluded).

In order to draw up this balance two simplifying assumptions were made:

a. technical relationships between production and abroad are the same as in the Netherlands (thus non-competing import is left out of the analysis);

b. Intermediate deliveries abroad follow the same patterns as in the Netherlands.

The figures of the balance of pollution, for certain pollutants are given in Table 14.

TABLE 14

DUTCH BALANCE OF POLLUTION (1973) FOR SOME POLLUTANTS AND FOR ENERGY CONSUMPTION

	poll. connected with export	poll. connected with import	balance	balance as % of nat. total
Total air pol. (APE)	24,483	14,751	9,732	23
Water poll. , BDOW (1000 p.e.)	11,411	8,897	2,514	10
Water poll. (HME)	251	276	-25	8
Energy consumption (PJ)	1,206	714	492	26

For most pollutants the balance is positive; in other words pollution is imported in the Netherlands. Further investigation pointed out that the sector "chemical basic products" was responsible for this to a large extent. When this sector is left out of the analysis, all above-mentioned balances, would become negative, except APE combustion emissions. In 1973 the Dutch balance of trade was also positive, so a positive balance of pollution is not amazing. However, if the sector "chemical basis products" is left out of the analysis, the sign of the balance of trade does not reverse.

If for each sector the results are analysed, it appears generally that (apart from "chemical basic products") those sectors with a positive balance of trade have lower emission coefficients, and those with a negative balance of trade have higher emission coefficients.

If the sector "chemical basic products" would not have disturbed this picture, it would be a good illustration of the theory of comparative advantages: environment being scarce, but productive capacity being sufficient in the Netherlands. It is true that "chemical basic products" is a sector with high revenues, but the expenses in the form of pollution appear to be high as well.

V. SCENARIO ANALYSIS, 1985

A. INTRODUCTION

Scenarios are descriptions of possible developments in the future; they are not hard predictions. A distinction can be made

* This sector includes "fertilizers" in this section.

between trand-scenarios, giving an impression of the future situation in the case of unchanged policy, and contrasting scenarios, depicting a future situation if certain policy targets changes were to be implemented. Our scenarios try to give an impression of the margins of policy. Therefore not absolute values but differences between the indicators in trandscenario and contrasting scenarios are of interest.

The scenarios relate to the year 1985. The trendscenario has been constructed on the basis of extrapolation, predictions and present policy intentions. For the economic sectors the interrelations in the input-output matrix were assumed to be the same as in the '73 coefficients. For the energy sectors and for emissions, new estimates of the coefficients in the 1985 were made.

By employing linear programming contrasting scenarios were built. Successively each of the targets has been optimized, while the other targets and the production levels of the sectors were restricted.

The variation of the contrasting scenarios when compared with the trendscenario reflects the pollution abatement to be achieved by structural changes, i.e., by changing the production levels of the economic sectors. Technological change, although it was estimated for '85, has been kept fixed, so pollution abatement to be achieved by technological changes was left out of the analysis. Therefore the factual margins of 24 pollution abatement are broader than our results indicate.

B. CONSTRUCTION OF THE TRENDSCENARIO

Production levels in 1970-1978 and predictions for 1980 were available at the aggregation level of 23 economic sectors. Graphical representation of this data indicated that linear extrapolation was suitable. The results have further been disaggregated and then supplemented by estimates of turnover in the energy sectors. In these last estimates time series, predictions, policy intentions and capacity-data of refineries were used. According to expectations regarding economic growth, which are more pessimistic now than some years ago, an average growth rate of 2.0% p.a. was implicitly assumed over the period 1973-1985.

Production levels have been checked and where necessary corrected by calculation of the resulting final demand.

C. TECHNOLOGICAL CHANGES

As for the interrelationships between economic sectors the 1973-input-output data has been assumed representative for 1985 also. Extrapolation of input-output coefficients might be possible (although it would be very difficult at our level of disaggregation), but would go beyond the scope of this research project. Only for the energy sectors new coefficients have been estimated because a policy aimed at conservation of Dutch natural gas resources is expected to lead to a fuel switch in favour of oil and coal. Large energy consuming utilities can be held to lead such a switch.

In addition, account has been taken of the possibilities for energy saving; some of these improvements were already realised in 1979, when this research project was finished.

Regarding emission coefficients considerable reductions of emissions per unit of production are being realised and are expected to continue to be realised until 1985. Therefore, new emission coefficients for the included pollutants were estimated. Similarly, the coefficients of employment-data have been corrected. Finally, two new sectors were introduced in the input-output matrix namely "water purification" and "treatment of solid (non-chemical) waste".

In this abridged report it has not been possible to present all the figures and their background. We refer to the original reports (Hordijk et al., 1979 a and b). In table 15 we give a summary of the situation in 1973 and the estimated situation as in the 1985-trendscenario.

TABLE 15
TRENDSCENARIO (1985) AS AGAINST THE YEAR 1973

	1973	1985
Energy consumption (PJ)	1,910	2,121
Air pollution (APE)	42,039	41,254
BDOW (1000 p.e.)	24,211	3,119
HME	327	112
Income (x million Dfl.)	120,061	157,694
Employment (x 1000 man years)	4,068	3,788

Due to further technological change it is expected that for the environmental indicators chosen here, the situation in the 1985-trend-scenario will not be worse than in 1973, as can be seen in Table 15.

D. CONTRASTING SCENARIOS

Contrasting scenarios have been constructed with the aid of linear programming. Seven criteria have been used: energy consumption (EC), air pollution (APE), water pollution by biologically degradable organic waste (BDOW) and by heavy metals (HME), and solid non chemical waste (SW); as economic indicators income (INC) and employment (EMPL) were chosen.

Constraints have been set to the production levels of the economic sectors: they were not allowed to deviate more than 15% from the levels in the trendscenario; this seemed to be reasonable margin over a period of five years.

Final demand has been kept non-negative. For some economic sectors ("retail trade", "real estate", "water purification", "treatment of solid waste", "other") final demand has been fixed on the level of the trendscenario; final demand of electricity has also been fixed and that of petrol, gas oil and fuel oil has been assigned positive lower boundaries. Moreover, it has been stipulated that for the seven criteria mentioned above the situation should not deteriorate, relative to the trendscenario.

Seven contrasting scenarios were built by optimizing each of the seven criteria subject to the above-mentioned constraints.

An eighth scenario was made by the use of a method of multi-objective programming which would allow each of the criteria to reach "good" values simultaneously.

The scenarios have been labelled with the following characters:

A: trendscenario
B: contrasting scenario by minimizing EC
C: contrasting scenario by minimizing APE
D: contrasting scenario by minimizing HME
E: contrasting scenario by minimizing SW
F: contrasting scenario by minimizing BDOW
G: contrasting scenario by maximizing INC
H: contrasting scenario by maximizing EMPL
I: contrasting scenario by multi-objective programming

The results of the scenarios with respect to the criteria are given in Table 16.

TABLE 16
EFFECTS OF NINE SCENARIOS (1985) ON SEVEN CRITERIA: VALUES EXPRESSED IN PERCENTAGES OF THE LEVELS IN THE TRENDSCENARIO

	A	B	C	D	E	F	G	H	I
EC	100	89.5	91.6	93.9	96.8	92.8	100	99.5	92.2
APE	100	92.7	89.6	93.1	92.5	92.0	100	100	92.1
HME	100	85.0	85.7	85.0	94.2	85.6	99.1	96.5	89.3
SW	100	100	93.9	93.0	92.2	100	100	100	95.5
BDOW	100	93.7	94.4	94.2	94.2	93.2	100	100	94.7
INC	100	100	100	100	100	100.2	109.0	108.9	102.9
EMPL	100	100.4	101.6	101.4	100	100	109.6	109.6	104.2

E. DISCUSSION OF THE RESULTS

In interpreting the scenarios not the absolute results but the differences and similarities of the scenarios are of interest. The restricting values, which were to some extent chosen arbitrarily, have a great influence on the absolute values of production levels. For instance, the constraints to the production levels of the sectors (deviation of the trendscenario of at most 15%) turned out to be almost always binding. Scenarios G and H prove to be very similar, so choosing income or employment as the main target gives about the same results.

Similarity between scenarios B and C could also be expected, because energy consumption leads to air pollution. They are different, however, which can be explained by the fact that energy is not only used for combustion but also as a raw material, and that process emissions are an important part of air pollution.

The contrasting scenarios indicate the desirability of expansion especially for services. For the sectors "banking, insurance", "business services" and "culture, recreation" the scenarios point in this direction without exception. The contrasting scenarios pointed also without exception towards expansion of the sector "livestock, extensive". However, no spatial constraint has been included in the model.

Consistency across scenarios with respect to reduced production levels was found for the sectors: "petroleum and coal products", "other chemical basic products" and "primary metals". For "fertilizers" only scenario E (minimizing solid waste) indicated expansion, all other scenarios indicated reduction. The production level of "paper" ought to be reduced except for the scenario H (maximizing employment) and "rubber and synthetic chemical products" should be reduced except for scenario G (maximizing income). For "gas" and "electricity", the indications for the production levels were not convincing, due to the strictness of constraints to energy sectors on which they depend strongly.

Sectors for which inconsistency was found between the scenarios aimed at economic targets (G and H) and environmental targets (B-F) were: "livestock, intensive", "meat products", "dairy products", "grain mill products", "margarine, oil, other foods", "floor coverings, other textiles", "lumber and wood products", "chemical final products", "fabricated metal products" and "electrical products"; the same tendency was found for "gas" and "electricity".

In all these sectors the environmental scenarios indicated reduction, the economic scenarios expansion of production levels. The multiobjective scenario mostly favoured the environmental scenarios. As could be expected, there exists a certain conflict between the environmental and the economic scenarios. In scenarios B-F the constraint to income is always binding to other words, improvement of the environment costs money. That the constraint to employment is not binding in scenarios B-F can be explained by the fact that income, which is strongly correlated to employment, was constrained. The improvement of employment relative to the trandscenario is very small.

It is noteworthy that in scenario B and F the constraint to non-chemical solid waste is binding. Thus, energy, consumption and BDOW could be less if more solid waste were accepted.

In the economic scenarios constraints to energy consumption, air pollution, solid waste and BDOW are binding or almost binding; only for water pollution by heavy metals a slight improvement with respect to the trendscenario can be observed.

Scenario I indictes that it is possible to attain an improvement for all criteria simultaneously: on average an improvement of about 10% of the ratio of environmental to economic criteria is possible if production levels of the sectors are not allowed to deviate more than 15% from the levels in the trendscenario.

When inspecting the production levels of the energy sectors, no great deviations from the trendscenario can be found for the sectors "gas" and "electricity". This is not surprising as final demand of these sectors was fixed at the level of the trendscenario. For "electricity" this concerns consumption by households, where no great deviations are to be expected. For natural gas, export is an important part of final demand, and most international contracts for delivery of gas are already concluded. Therefore, it did not seem realistic to allow any great deviation from the trendscenario.

Important changes are found in "petroleum and coal products". These concern changes in the energy mix: the reduction in the production levels of gasoil, naphta, light and heavy oil is such that exports disappear; petrol and dieseloil however, expand considerably.

There will be a connection between these movements and the prices of these types of energy. It should be remarked here, that there were no constraints in the model with respect to foreign trade (neither constraints to the product mix of the refineries).

F. FINAL REMARKS

The improvements which can be attained by structural abatement seem to be less than in a proceeding study (Jansen et al., 1978). This may be explained by the more stringent constraints used in this report. In this scenario analysis, the numerical results are to a large extent dependent on the to some unavoidable extent, arbitrarily chosen restricting values. The relevant information however, lies primarily in the directions to which the production levels of the various sectors should develop.

Two points remain to be mentioned. In the first place it must be kept in mind that structural abatement is just one method to keep pollution generated by productive activities within limits. Technological improvements are a complementary method. The technological changes to be expected between 1973 and 1985 are turning out to be promising. If variations with respect to technological processes were allowed in the model, we expect that the results would be much better.

In the second place we want to stress that in this study only pollution generated by productive activities was considered. Consumptive activities have a polluting impact as well (Vos and Olsthoorn, 1979), and for these activities pollution abatement policy can also contribute to a reduced pressure on environment.

In conclusion, notwithstanding strong doubts whether in the long run continued economic growth is compatible with an acceptable quality of the natural environment, we think that there are at the moment opportunities to fruitfully contribute to the amelioration of the pollution problem.

APPENDIX A

CLASSIFICATION OF 60 ECONOMIC SECTORS

1. Livestock (extensive)
2. Livestock (intensive)
3. Horticulture
4. Other agriculture
5. Fishing
6. Coal mining
7. Pegroleum and natural gas
8. Other mining
9. Meat products
10. Dairy products
11. Canning, preserving
12. Grain mill products
13. Sugar
14. Bakery products
15. Confectionery products
16. Margarine, oil, other foods
17. Beverages
18. Tobacco
19. Spinning
20. Weaving
21. Knitting
22. Floor coverings, other textiles
23. Clothing
24. Leather, shoes
25. Lumber and wood products
26. Paper
27. Paper products
28. Printing and publishing
29. Petroleum and coal products
30. Fertilizers
31. Other chemical basic products
32. Chemical final products
33. Rubber and synthetic chemical products
34. Building materials
35. Primary metals
36. Fabricated metal products
37. Machinery
38. Electrical products
39. Automobile manufacture
40. Other transportation equipment
41. Other manufacturing
42. Electric utilities
43. Gas
44. Water
45. Construction
46. Wholesale trade
47. Retail trade
48. Hotels, restaurants, bars
49. Maintenance and repair
50. Sea and transportation
51. Other transportation
52. Communications
53. Banking, insurance
54. Real estate
55. Business services
56. Social services
57. Medical services
58. Culture, recreation
59. Other services
60. Other

In the scenario-analysis for 1985 the sectors: "water purification" and "treatment of solid non-chemical waste" are also included.

BIBLIOGRAPHY

Central Bureau of Statistics, 'Waterkwaliteitsbeheer', The Hague, (1978).

Dorfman, R. et al., Linear Programming Analysis, McGraw Hill, New York, (1958).

Hartog and Houweling, 'Economic Consequences of Pollution Abatement', Central Planning Bureau, Monograph no. 20, The Hague, (1975).

Hordijk, L., et al. (a), 'Economische Structuur en Milieu. Productie, Milieuverontreiniging en Energieverbruik 1973/1985, VAR 1979/7, Ministry for Public Health and Environmental Affairs, Leidschendam, (1979).

Hordijk, L. et al. (b), 'Milieuverontreiniging en Productiestructuur in Nederland, Deel iii.' Institute for Environmental Studies no. 79/6, Free University, Amsterdam, (1979).

James, D., et al., Economic Approaches to Environmental Problems. Elsevier, Amsterdam, (1978).

Jansen, H.M.A., et al., 'Environmental Pollution and the Economic Structure in the Netherlands,' Institute for Environmental Studies, Free University, Amsterdam, (1978).

Leontief, W., The Structure of the American Economy 1919-1939. Oxford University Press, Oxford, (1951).

Metra Consulting, 'Long Term Consequences of NO_x and SO_2 Abatement', London, (1978).

Ministry for Economic Affairs, 'Economische Structuurnota', The Hague, (1976).

Ministry for Public Health and Environmental Affairs, 'Bestrijding van de Luchtverontreiniging, Indicatief Meerjarenprogramma 1976-1980', The Hague, (1977).

Olsthoorn, A.A., and J.B. Vos, 'Milieuverontreiniging en Consumptieve Activiteiten, Interimnota', Institute for Environmental Studies, Free University, Amsterdam, (1979).

Tinbergen, W. and P. Schake, 'Syllabus Inleiding Milieukunde,' Technological University, Delft, (1977).

Victor, P., Pollution, Economy and the Environment, Allen and Unwin, London, (1972).

Vos, J.B., 'Internationale Handel en Milieuverontreiniging: de Vervuilingsbalans', Ec. Stat. Berichten, 20/27-12-'78, (1978).

Werkgroep NO_x-vorming Vuurhaarden, 'Het Verminderen van de NO_x-uitworp van Vuurhaarden, VAR 1978-6'. Ministry for Public Health and Environmental Affairs, Leidschendam, (1978).

III Systems and Models for Energy Environmental Assessment in the U.S.

10 An Overview of the Strategic Environmental Assessment System

SAMUEL RATICK AND T. R. LAKSHAMANAN

I. INTRODUCTION

During the last decade we have witnessed a growing concern over the long term physical capacity of the earth to provide for our expanding natural resource needs and to meet an desire for a high amenity environment. This increasing concern, however, has not been matched by an improved ability to incorporate natural resource and environmental imperatives in societal decision-making. This has resulted in conflicting and inconsistent recommendations often being made regarding long term strategic directions. While the disparities between long term and short run objectives in modern day democratic societies are in part responsible for these inconsistencies, a major cause for this problem is our inability to assess the long term effects of the complex interactions among economic activity, natural resource usage and environmental quality.

An increased understanding of these complex interactions and their implications for long term planning options is necessary in order to:

- identify future development constraints and the fundamental societal choices that need to be made for release from these constraints;
- develop an interim strategy towards the environment and natural resources that retains desirable policy outcomes as progress towards long term goals is made, and
- encourage an increased awareness of multiple goals within a planning framework that reduces goal conflicts and increases policy coherence.

Such knowledge and conscious long range social decision-making is required for the continuation of a high amenity industrial society assured of ample supplies of moderate cost energy, raw materials and a healthy environment.

In recent years several analytical efforts have been made to characterize future resource and environmental problems in a quantitative fashion. From the introduction of the "World Models" developed for the Club of Rome, more than a dozen models have been developed to address the resource and environmental futures in developed and developing nations (Table 1). These models vary in functional scope (e.g., a single sector such as energy or multisectoral), in geographical dimensions (supranational, national,

SCOPE / GEOGRAPHIC SCALE	COMPREHENSIVE (MULTISECTORAL)	SINGLE SECTOR
SUPRANATIONAL	World 2 World 3 Leontief et al Mesarovic-Pestel LINK Bariloche Japanese Club of Rome	Queen Mary College Model Nordhaus Demand Model EURMAP Nordhaus Resource Allocation Model Haefele-Manne Transition Model
NATIONAL	SEAS DRI (LITM)-BNL (DESOM) Lesuis-Muller-Nijkamp Model	
SUBNATIONAL	Delaware Estuary Model Intermedia Flow Model Mesophytic Forest Model	Agricultural Runoff Management Model Livermore Regional Air Quality Model

Table 1. Types of Strategic Analysis Models.

and regional), in time scale (20-100 years) and in solution framework (predictive or normative).

The earlier strategic models typically were comprehensive in scope, highly simplified and very aggregative in specification and poor in data as exemplified by the Latin American World Model and the Bariloche model. For the Bariloche Model six sectors (population, health, food, education, nonresearch resources, energy and pollution) and four regions characterize the world (Figure 1). Capital and labor are allocated within a linear programming framework to meet each individuals basic minimum needs of food, housing, health and education subject to certain constraints, the objective function is to maximize life expectancy at birth.

While these earlier strategic models have played an important role in spotlighting the important interactions among the economy, natural resource usage (energy in particular) and environmental pollution, they are far too aggregative in functional scope, time scale and spatial detail and too data poor to have specific policy value.

It is in this regard that the second generation strategic models such as the Strategic Environmental Assessment System (SEAS), DRI and DESOM models are relevant. These models deal with one country, disaggregate their scope into a very large number of economic sectors, energy end uses and environmental sectors. They focus on the intermediate run (10-25 years). They deal with the effects of changes in income, technology (of production, pollution control and energy extraction) and consumption. Then developers have invested heavily in data acquisition and frequent updating of data bases. The model users deal with uncertainties and complexity by use of the scenarios and series of special studies.

This paper describes the SEAS model and its policy analysis applications in the U.S. Federal Government. The SEAS model was originally developed in the U.S. Environmental Protection Agency for assessing future pollution emissions consequent on differing growth scenarios and the concomitant cost of pollution abatement over a 15-20 year horizon (Lakshmanan and Krishnamurthi 1973). It subsequently has been expanded in scope (House 1976, Lakshmanan and Ratick 1977, Ridker and Watson 1979). The following sections of this paper and the three subsequent papers describe the SEAS system and its policy applications.

II. DESCRIPTION OF THE SEAS SYSTEM

The Strategic Environmental Assessment System (SEAS) is an integrated series of computer implemented models and data bases. Its purpose is to provide a systematic analytical framework for conditional analysis describing how economic and demographic activities affect and are dependent upon the physical envirnment. Figure 2 is a schematic of the SEAS depicting the relationships among the substantive areas that are included in the system.

SEAS provides medium-term forecasts (15 to 20 years) of economic activity and attendant resource usage and environmental pollution based upon input assumptions such as economic, population and energy demand projections. Each module is sufficiently detailed to provide specific

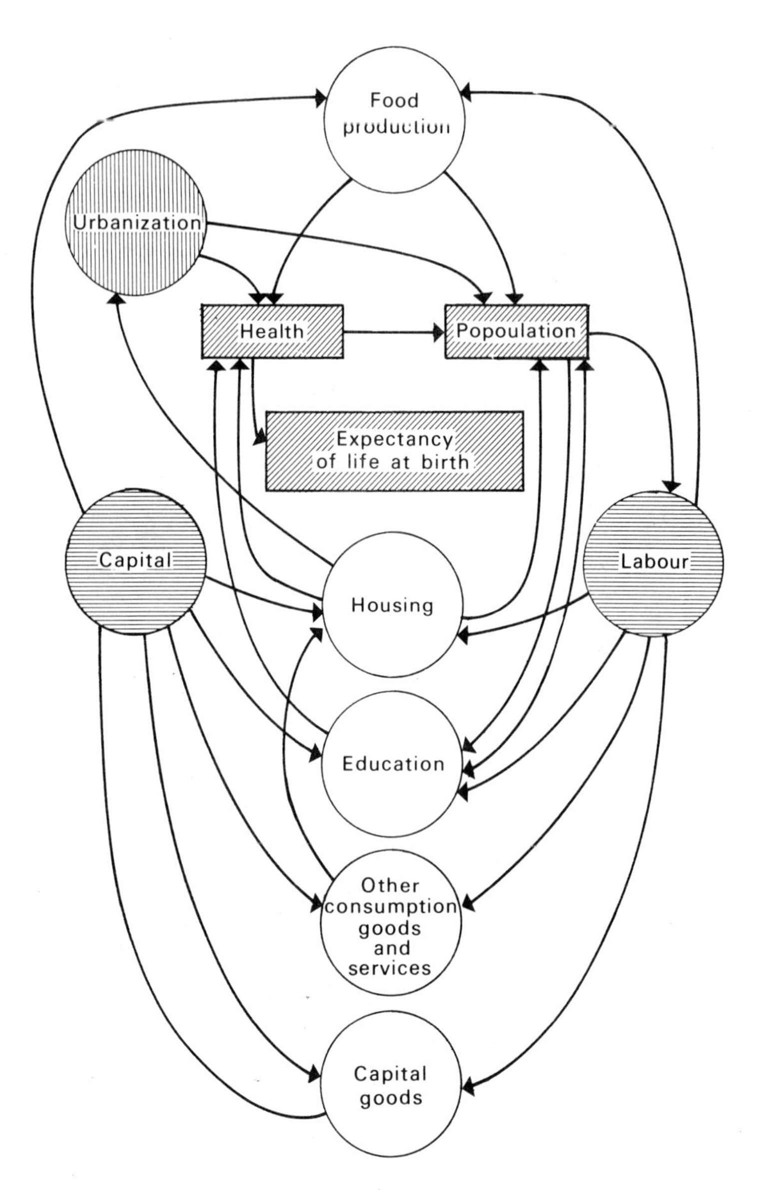

Figure 1. The Basic Structure of a Block in the Bariloche Model.

information on the sectoral composition and spatial distribution of the outputs of these forecasts.

For expository purposes the system can be partitioned into three functional areas, economic, energy and environment. Each of these partitions is made up of many modules designed for a specific purpose that are interrelated within and between partitions by functional relationships and data matrices. The following is a brief description of the important modules within each of the partitions.[1]

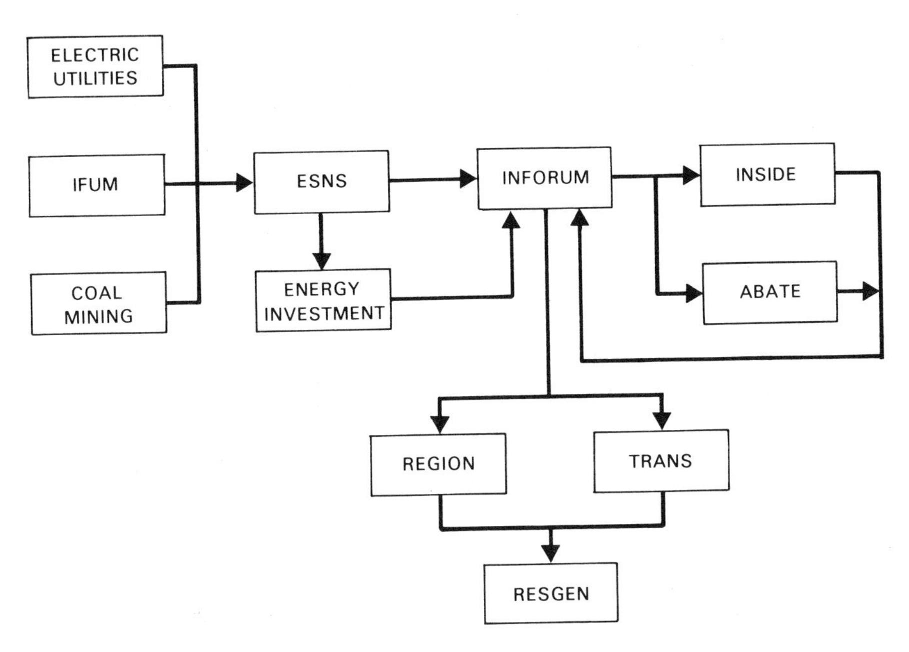

Figure 2. The Basic SEAS Modules.

Economic

Central to this partition is SEAS/INFORUM[2], a modified 200 sector dynamic input/output forecasting model. SEAS/INFORUM provides year by year forecasts of activity levels for each of the 200 economic sectors in response to macroeconomic projections of final demand. An input/

---

1 For a more detailed SEAS description see House (1976), Lakshmanan and Ratick (1977, 1980). "A Guide to SEAS Documentation" prepared for the U.S. Department of Energy by the MITRE Corporation is a compendium of available SEAS documentation.

2 Interindustry Forecasting Model of the University of Maryland (INFORUM) developed and maintained by Professor Clopper Almon and his associates. Almon et.al. (1974).

output framework is used because of its detailed and complete description of the U.S. economy and its flexibility in allowing new sectors to be incorporated. Standard matrix solution procedures simplify the calculation of direct and indirect economic effects resulting from changes in final demand due to differing exogenously supplied input assumptions.

For many energy and environmental analyses, 200 sector detail is not sufficient, therefore, many economic sectors have been disaggregated into subsectors. Sector disaggregation is based primarily upon the varied physical products produced within an economic sector or upon differing technological processes used to produce a particular product. Algorithms of the INSIDE module compute the subsector activity levels. Figure 3 is a schematic of the subsector disaggregation process. Currently, the INSIDE module contains over 350 subsectors of product and process categories. Information provided by the computational routines of this module include the physical amount of product produced by process and by industry. Thus, for economic sectors important to energy and environmental analysis, economic activity levels are closely tied to detailed process flows and product outputs.

Energy

The modules of SEAS related to energy systematically calculate the demand for energy, the supply patterns needed to meet that demand, and the capital needed to finance new suppply activities. Figure 4 shows the relationship of the energy partition to other modules. The demand for energy can either be exogenously input or derived within the three demand modules, Residential/Commercial, Industrial and Transportation, using the economic activity levels from INFORUM. This yearly forecast of energy demand is input to the Energy System Network Simulator (ESNS),[3] a network flow model, that calculates the related energy resource consumption and energy supply technology activity levels.

As with most other SEAS modules ESNS is transparent allowing for many user supplied changes to be incorporated. Some of these input changes include: allocation between competing energy supply technologies (e.g., solar, geothermal, fossil fuel and nuclear technologies for electric generation), specification of net energy efficiencies for each technology and supply constraints for each link in the network.

The calculated activity levels for energy supply technologies are input to the Energy Investment Module which estimates the investment pattern needed to support these levels. This investment requirement is "fed back" to SEAS/INFORUM and to the energy demand modules (Figure 5).

Modules that calculate regional energy consumption or supply patterns will be discussed in the next section.

---

3 Developed at Brookhaven National Laboratories, see Hoffman and Palmedo (1972), and Placet and Heller (1979).

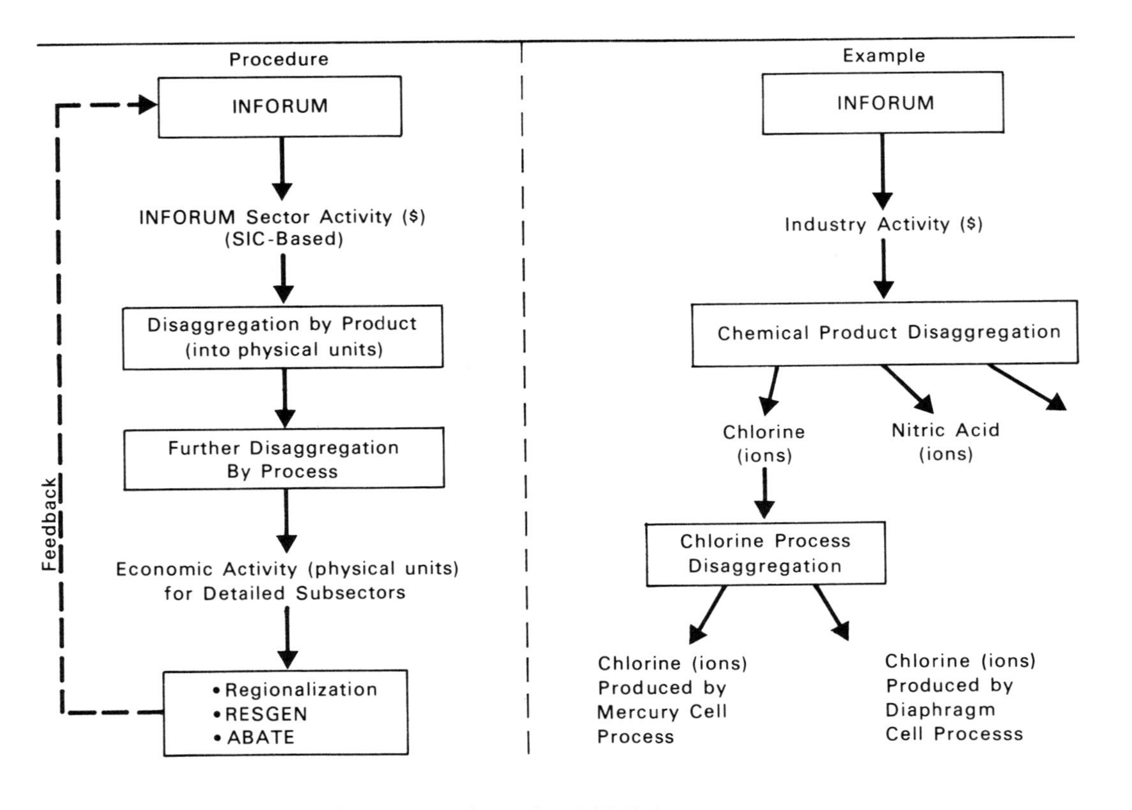

Figure 3. Inside Module Dissaggregation of INFORUM Sectors.

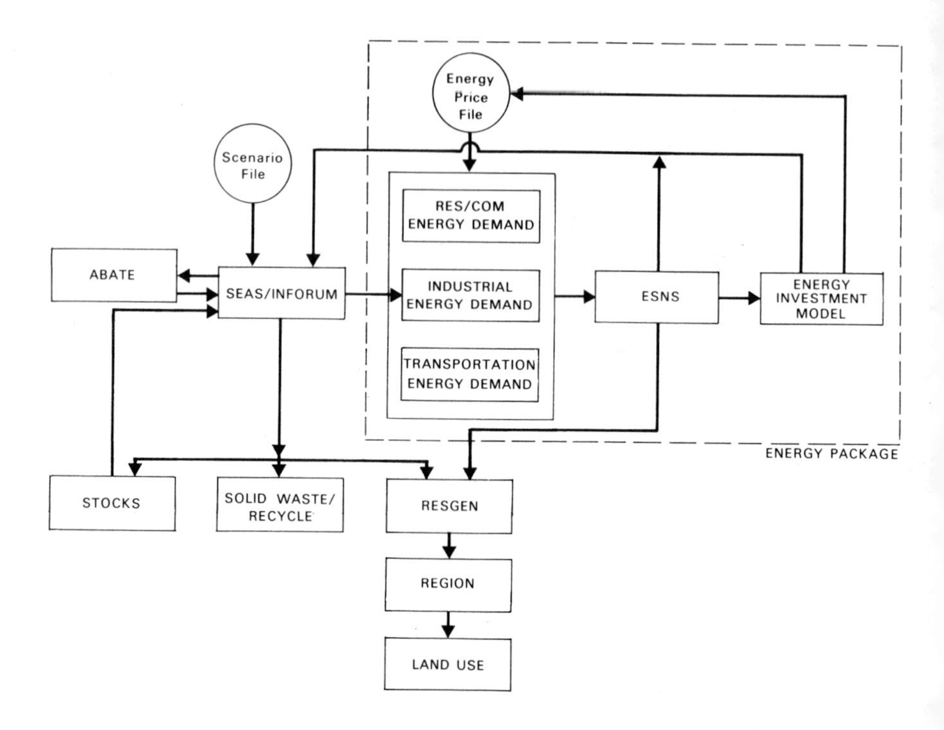

Figure 4. SEAS/TRANS Module Flow.

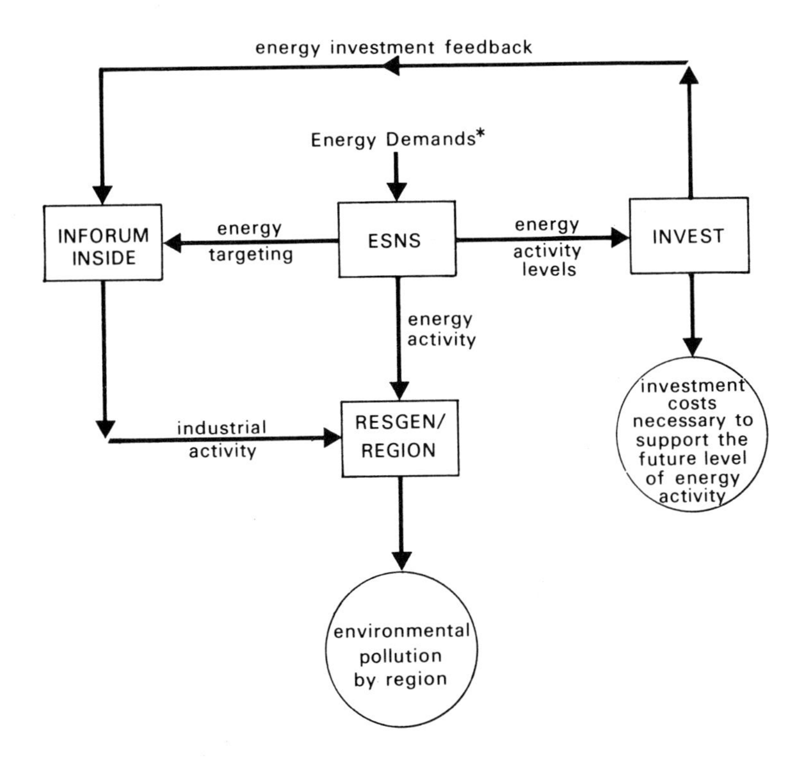

**These energy demands are either calculated by three demand modules (transportation, residential/commercial and industrial), which are driven by INFORUM or input from an exogenous source such as MEFS.*

Figure 5. ESNS Module Interactions.

Environment

SEAS was developed to help decision makers in the Environmental Protection Agency (EPA) anticipate environmental problems. Although the outputs of the modules in the economic and energy partitions are useful and informative, the primary purpose for calculating this information in such a consistent manner is to provide decisionmakers with accurate and comprehensive forecasts of pollution generation, ambient environmental quality and environmental costs and benefits.[4] There has been significant module development aimed at providing forecasts of pollution generation. However the modules that link these generated levels of pollution to measures of ambient environmental quality and to environmental costs and benefits are not well developed within the SEAS framework and will not be discussed here.

The modules in the environment partition relate the level of activity calculated by the modules of the economic and energy partition to concomitant levels of residual (pollution) generation. The investment and operation and maintenance costs for exogenously supplied levels of pollution abatement activity, are calculated and "fed back" to modules in the economic partition. Thus, the amount and composition of pollutants that will be generated by economic or energy related activity levels forecast by SEAS, is tied to the specific technologies employed for those activities and to the type and degree of abatement.

Calculation of the total amount of pollution generated nationally has only limited usefulness. It is therefore necessary to not only relate levels of economic activity to specific technological processes but to locate these processes spatially within regions of the U.S. The linking of economic and demographic activity to detailed technological processes is accomplished within the INSIDE module of the economic partition and by ESNS of the energy partition. The REGION module allocates these activity levels to regions.[5]

REGION converts information on a national scale into forecasts for federal regions, states, standard metropolitan statistical areas (SMSA), air quality control regions (AQCR), hydrologic aggregated sub areas (ASA) and major and minor river basins.

Regionalization is accomplished by several different methods applied for different technologies. A shift/share methodology, uses data on current regional locations of activities and estimates of future shifts in composition. National economic activity is allocated to regions each year of the forecast based upon that region's calculated share of each economic sector. A second methodology takes advantage of the "transparency" of the modules by allowing for overrides where more detailed and specialized regional data is available. The third method uses actual locations of existing and

---

[4] For further information on the methodology for linking generation and ambient air quality and benefits see Ridker and Watson, (1979).

[5] For a critical analysis of the Region Module see Lakshmanan et al. (1979).

proposed facilities or technologies; these "siting" methods of REGION are specified for industrial fuel use, electric utilities and coal mines.

The linking of activity levels and the generation of residuals (environmental pollutants) is accomplished within the RESGEN module, Figure 6. A residuals coefficient matrix relates activity levels to gross residual generation, Table 2. Under user specified or programmed default assumptions about the level and type of abatement activity, net, captured and secondary residuals are calculated. Net residuals are those released to the environment. Secondary residuals are those captured by the abatement process that are transformed to another form, for example air pollutants that have become a solid or liquid waste such as scrubber sludge. The output of this sector is quite detailed providing information on pollution generation by category of pollutant (e.g., particulates), pollution component (e.g., particulate lead) or by economic or energy sectors (e.g., chlorine production or electric generation). This output is provided for each forecast year and can be obtained under changing assumptions of scheduling and degree of abatement. This information can be calculated on a national level or region by region.

Mobile source pollution is calculated at the regional level within the Transportation module in a generically similar manner. Pollution coefficients are given by model year based upon user input control strategy assumptions. Each forecast year the original pollution coefficients for that model year are increased based upon a deterioration factor representing the loss of efficiency of the inplace abatement technology with age. Pollution generated from transportation activities is a function of vehicle age and the miles traveled by vehicle type.

The relationship of the RESGEN module to REGION is shown in Figure 7.

III. APPLICATIONS OF THE SYSTEM

SEAS is operationalized using exogenously supplied input scenario assumptions and user input changes. In the extreme every data element, parameter and algorithm can be changed for any given run. However, the system has been designed to accept certain changes quite readily. Macro scale inputs such as Gross National Product (GNP) forecasts or end-use energy consumption by fuel type are easily input. Microscale changes in each module are also accommodated in the system. For example, in the RESGEN module the user can specify, at the beginning of the run, the years that specific industries will meet prespecified environmental standards. The routine in the module then interpolates the efficiency of removal by pollutant and by technological sector (or process) for each intervening year. The level of removal of pollutants can also be modified simulating changes in environmental policy such as more or less stringent emission and effluent standards.

Although default parameters values are given in each module, the user of the system must be very careful to ensure the macro scale system inputs and micro scale changes in each module are consistent. This "transparency" of the system allows SEAS to be used for many types

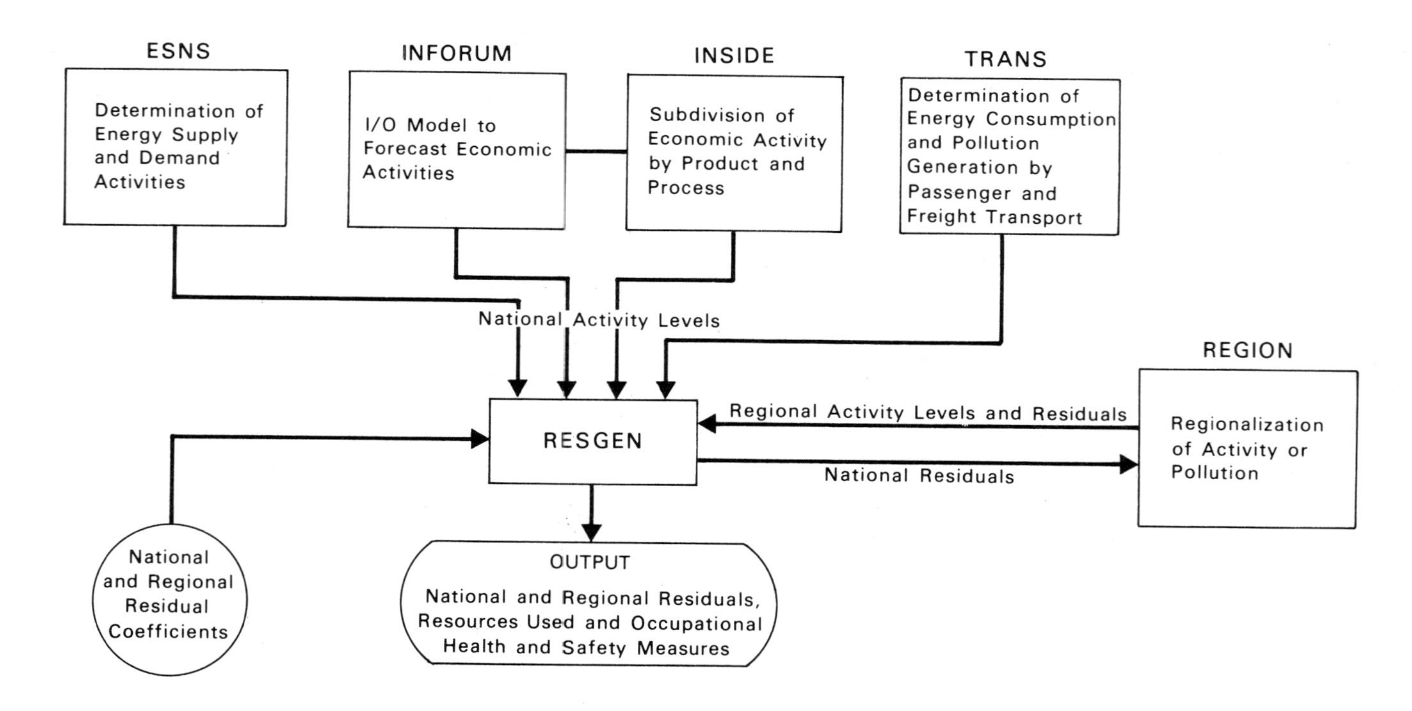

Figure 6. The Relationship of Resgen with Other SEAS Modules

TABLE 2

RESGEN METHODOLOGY FOR GROSS RESIDUALS

Activity Level in a target year	X	RESGEN COEFFICIENT for the taget year	=	GROSS RESIDUAL (resource consumed) in the target year
. 10^{12} Btu input		. Tons per 10^{12} Btu input		. Tons
		. Person-days per 10^{12} Btu input		. Person-days lost
		. Acre-feet of water per 10^{12} Btu		. Acre-feet of water used
. Million $ output		. Tons per million dollar output		
. Million tons output		. Tons per million tons output		

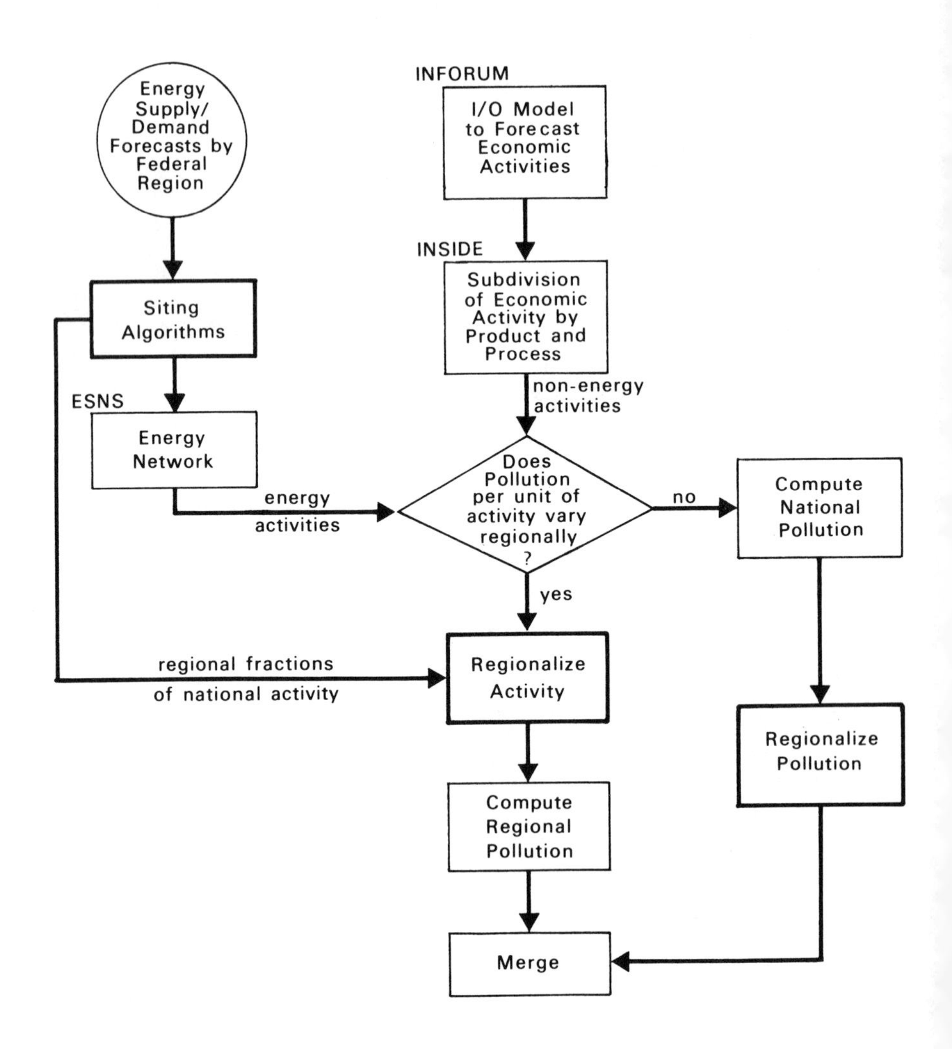

Figure 7. The Relationship Between Regionalization and Other SEAS Functions.

of applications and in many different ways. It also restrict how the system is used and by whom.

The complexity and flexibility of the SEAS limits the running of the system to teams of analysts and system specialists, however, specification of important input changes and some analysis of output can be done by analysts other than those who run the system. Currently the two major users of SEAS are the U.S. Environmental Protection Agency (EPA) and the U.S. Department of Energy (DOE). Both agencies have contracted to teams of analysts whose primary function is to run the system. SEAS was also used by other government agencies and private institutions. Notably, Resources for the Future (RFF) used a modified version of SEAS (SEAS/RFF) to "assess the resource and environmental problems that the United States may face over the next fifty years" (Ridker and Watson, 1980).

Any use classification for such a comprehensive general purpose forecasting system will neither be exhaustive of types of uses nor provide mutually exclusive categories for classification. At best such a scheme can only suggest some ways in which the system may be utilized. With this caveat in mind, we offer the scheme exhibited in Table 3 as one way of classifying the uses for SEAS.

The systems applications are organized according to the purpose for the application and the way the system was or will be applied in the analysis. The entire system may be operationalized or just selected modules within each of the economic, energy and environmental partitions. The data contained in any of the modules of the system can be used alone without any of the system algorithms being run. SEAS has and can be used as the principal technique for analysis or applied to provide data and information to be used in conjunction with other analytical methods.

The system is used primarily to perform conditional analyses of the impacts of alternative environmental or energy strategies, which is usually done by comparing the results of different scenario runs. A reference or "base case" scenario, that simulates current conditions or a "business as usual" projection, is used as the basis of comparison. Alternative scenarios, simulating the proposed strategy, are run and their results compared to the reference scenario. This "with and without" comparison provides a measure of the impacts of the action along several dimensions; a comparison of levels of effects, their sectoral composition and the spatial distribution of impacts for each year over the forecast horizon.

SEAS can provide useful information and analyses for public sector planning activities that have to allocate resources to competing research or implementation programs. Such programs may be aimed at mitigating adverse environmental conditions or developing and promoting environmentally acceptable energy supply technologies. SEAS can also be used to aid in determining the consequences of various energy and environmental policies by projecting their spatial and sectoral impacts. Ways in which the policy may be modified to either eliminate unwanted consequences or augment desirable effects can thus be obtained.

Another application of SEAS is to provide a framework for the

TABLE 3. CLASSIFICATION OF SEAS APPLICATIONS

The Purpose for the Application	The Way the System is Applied	
	Degree of Utilization	Extent of Involvement
* Planning	* Entire System	* Principal Technique
* Policy Impact Analysis	* Selected Modules	* Contributing to Analysis
* Integrative Tool for Analysis		
* Informational Projections and Forecasts	* Data Bases and Elements	

integration of disparate parts of a large study. The systems structure could provide an organizing theme for data collection and analysis and also insure that the various tasks of the study are coordinated and built upon a consistent set of assumptions.

Table 4 lists some representative applications of SEAS. The words listed with each application correspond to the underlined descriptors in the taxonomy.

As an example of the types of information that can be obtained from the system, representative results from the first application listed in Table 4 will be presented. SEAS was used to assess the environmental and economic implications of the National Energy Plan (NEP) proposed by President Carter April 29, 1977. Almost all modules were used and the SEAS was the principal analytical method employed. The conditional analysis was performed by calibrating the various modules of the system to the energy and environmental control initiatives and economic assumptions of the NEP.[6] The results of NEP system run were compared to a "base case" (called the Pre-NEP scenario) made without the NEP initiatives. The comparison of the results of the two scenarios provided detailed information on the overall levels of impact, the effects on specific industries or resources and the spatial distribution of these effects.

Table 5 is an example of the kind of output available from the analysis displaying the overall level of effect. The five water pollutants listed show differential growth rates over the forecast horizon and different lists of major contributors. Total dissolved solids shows the largest growth rate over the 25 years due primarily to growth in energy related activities.

Comparisons can also be made on a sector-by-sector basis as shown in Table 6 for projected emissions of the five criteria air pollutants. Particulates and sulfur oxides are primarily from "fixed" sources (industries and utilities) while hydrocarbons and carbon monoxide are the result of mobile source emissions (cars and trucks). For all sources of air pollution except industrial combustion of gas and coal the NEP scenario produced less air emissions in 2000 than the Pre-NEP scenario. This is due to the stricter environmental standards proposed in the NEP.

An example of the spatial distribution of effects for emissions of sulfur oxides by federal region is given in Figure 8 and 9. By the year 2000 Federal Regions 4, 5 and 6 will show increases in SO_x emissions while overall emissions from Region 3 are relatively constant over the 25 year forecast. Under each scenario and for each year of the forecast Region 5 emits the largest amounts of sulfur oxides.

---

6 See "Environmental Challenges of the President's Energy Plan: Implications for Research and Development," report prepared for the Subcommittee on the Environment and the Atmosphere of the Committee of Science and Technology, U.S. House of Representatives by the Congressional Research Service, Library of Congress, GPO No. 052-070-04274-5, October 1977.

TABLE 4

EXAMPLES OF SEAS APPLICATIONS

- Assessing the Environmental Impact of the President's National Energy Plan (U.S.).
  Policy, Entire, Principal

- Transportation Energy Conservation Network (TECHNET).
  Policy, Modules, Contributing

- Integrated Technology Assessments (ITS)
  - Ohio River Basin Study (ORBES)
    Integrative (Planning), Entire, Principal

  - Western States (ITA)
    Planning, Entire, Contributing

- Resources for the Future (RFF) Study
  Integrative (Informational), Entire, Principal

- Offshore Oil and Gas Leasing (Bureau of Land Management)
  Policy, Data Bases (Modules), Contributing

- Annual Environmental Analysis Report (U.S.D.O.E.)
  Planning, Entire, Principal

- National Environmental Impact Projections (U.S.D.O.E.)
  Policy, Entire, Principal

- Five Year Outlook (U.S.E.P.A.)
  Integrative, Entire, Principal

- The Council on Environmental Quality (CEQ) Annual Report
  Informational, Entire, Contributing

- National Commission on Water Quality
  Policy, Entire, Contributing

- Cost of Clean Air and Water Report (1977) (U.S.E.P.A.)
  Integrative, Entire, Principal
  Modules, Contributing (1978....

- Assessing Solar Energy Growth Scenarios
  Planning (Informational), Entire, Principal

- Employment Impacts of Environmental Laws (Bureau of Labor Statistics)
  Policy, Entire, Contributing

TABLE 5

NET DISCHARGES AND SOURCES OF SELECTED WATER RESIDUALS

IN 1975, 1985 and 2000 UNDER THE NEP SCENARIO

(In Thousands of Tons)

Water Residual	1975	1985	2000	Ratio of 2000 Level to 1975 Level
Biochemical Oxygen Demand	3,721	1,377	1,760	47%
Total Dissolved Solids	13,341	10,142	14,234	107%
Suspended Solids	14,684	1,466	1,800	12%
Nutrients	1,130	713	939	83%
Oils and Greases	597	242	242	41%

The scenario runs for this conditional analysis were part of an ongoing effort in the U.S. Department of Energy (D.O.E.) which produced the Annual Environmental Analysis Report (AEAR) to provide information to the planning division of the D.O.E. on the likely environmental consequences of energy activities.

Subsequent efforts evolved from the AEAR activity such as the National Environmental Impact Projection series that used SEAS to provide impact assessments of differing scenarios provided by the Energy Information Agency.

As mentioned previously, pollution emissions could be regionalized to air quality control regions (AQCR). Figure 9 taken from the National Environmental Impact Projections No. 1, is an example of the spatial detail available from SEAS analysis.

Other applications of the system range from the use of specific data bases and modules for analysis to using the results of scenario runs requiring the entire system.

The use of SEAS by the U.S. Department of Interior's Bureau of Land Management to assess the impacts resulting from activities associated with offshore oil and gas exploration and development in the South Atlantic region initially utilized just the RESGEN data bases. The output of a regional econometric model was used to input economic activity levels to the RESGEN data matrices. Later applications of SEAS to offshore leasing studies used more of the environmental partitions data and module routines.

TABLE 6

PROFILE OF THE PROJECTED CRITERIA AIR EMISSIONS FOR THE YEAR 2000: NEP VS. PRE-NEP (10^3 Tons)

	Particulates			Sulfur Oxide			Nitrogen Oxide			Hydrocarbons			Carbon Monoxide		
	PRE-NEP	NEP	Diff*	PRE-NEP	NEP	Diff.	PRE-NEP	NEP	Diff.	PRE-NEP	NEP	Diff.	PRE-NEP	NEP	Diff.
Petroleum Refining	219	147	72	1968	1317	651	905	605	300	709	486	223	70	47	23
Coal Utilities	1342	1098	244	17450	14770	2680	7908	7180	728	135	119	16	455	399	56
Oil Utilities	23	15	8	1037	677	357	676	442	234	6	4	2	32	21	11
Gas Utilities	--	--	--	--	--	--	139	63	76	--	--	--	--	--	--
Coal Combustion (Ex. Utilities)	957	764	193	6743	8406	(1663)	2206	3965	(1759)	147	273	(126)	249	546	(297)
Oil Combustion (Ex. Utilities)	655	194	461	3315	915	2400	1441	410	1031	115	39	76	162	58	104
Gas Combustion (Ex. Utilities)	--	--	--	--	--	--	832	851	(19)	24	35	(9)	103	118	(15)
Trucks	600	580	20	587	571	16	6930	6735	195	4701	4569	132	37400	36378	1022
Auto	1439	1021	418	346	245	101	3043	2912	131	2563	2453	110	10307	9867	440
Other	8765	8681	84	2254	2400	(146)	4920	4837	83	3900	3422	478	10522	9966	556
TOTAL	14000	12500	1500	33700	29300	4400	29000	28000	1000	12300	11400	900	59300	57400	1900

*Difference (i.e., PRE-NEP total minus NEP total).

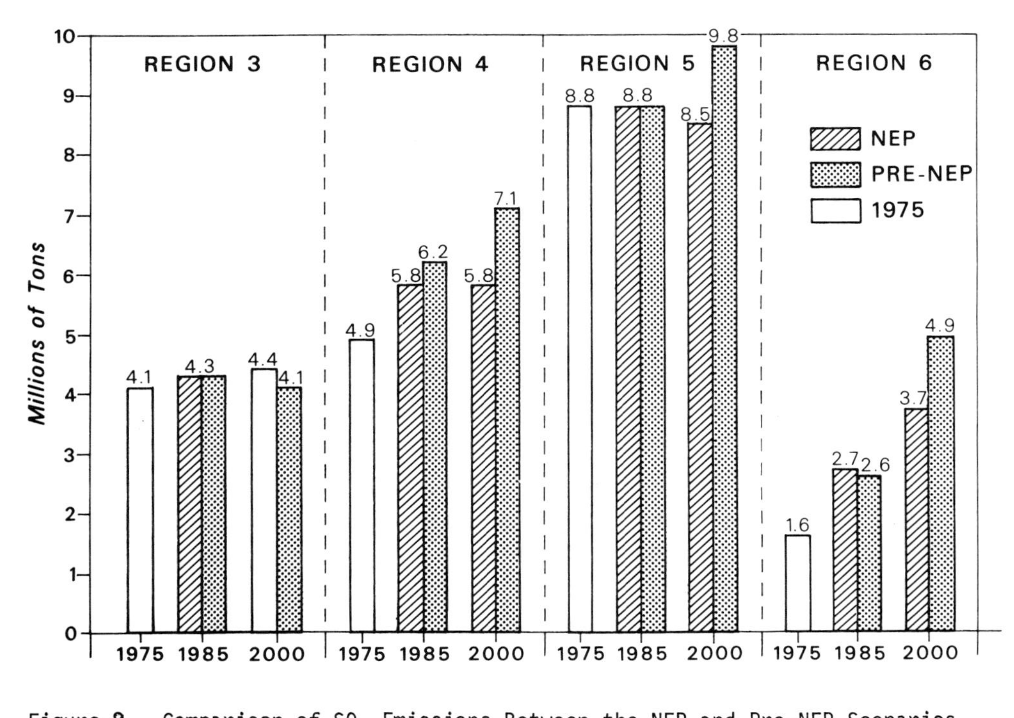

Figure 8. Comparison of SO_x Emissions Between the NEP and Pre-NEP Scenarios (for selected regions)

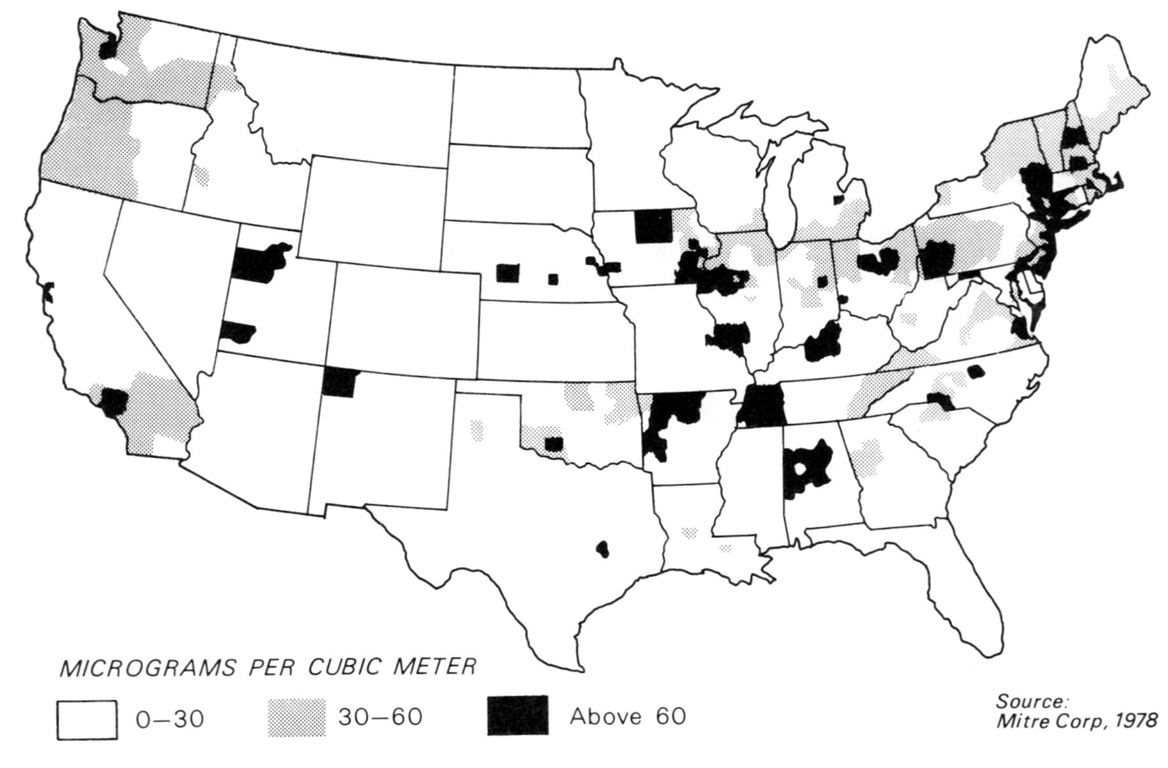

Figure 9. Projected Average Annual Ambient Sulfur Dioxide Concentrations in 1990 by AQCR.

The Office of Research and Development (OR&D) of the Environmental Protection Agency (EPA) has used the system as an integrative tool to aid in anticipating future environmental problems. SEAS was used to produce results that were extended, supplemented and evaluated by other sources and experts throughout the U.S.

IV. SUMMARY AND CONCLUSION

The Strategic Environmental Assessment System is a large, flexible transparent and in many ways complex predictive tool that has been applied in many different situations and for a wide variety of different purposes. These attributes of the system have provided many benefits to its users. Due to its scope the system provides a comprehensive view of the interactions among economic, energy and environmental activities. The integrated nature of the system assures a high degree of consistency in the data used, assumptions made and algorithms utilized to make its projections. The flexibility of the system and transparency to changes in the data and procedures in its modules allow its analytical capabilities to be specifically tailored to problems. Its large and complex structure permits a sufficient level of detail to be achieved in many policy applications.

The paradox such systems pose is that the attributes which enhance their usefullness can also plague their acceptance. The scope and complexity of SEAS makes understanding the entire system difficult for those who are to use the system and who must trust the information the system provides. Continued use and successful application of the system can provide an informal "in vivo" system verification. However, users often, quite understandably, require that such a system must first be verified before they will use it. Familiarity, in these circumstances, will breed a guarded acceptance.

Misunderstanding of the intent of the system or what kinds of information it does provide can render its analysis capabilities too specific for some uses and too general for others. Proper understanding of the level of geographical and sectoral aggregation designed into the system is essential to its successful application.

The system is capable of providing large amounts of useful information in a relatively short period of time. To do so severe demands are placed upon the systems specialists to coordinate the different tasks necessary for implementation of the analysis. Underestimating the time it takes for a complete and "debugged" analysis by both the systems analysts and users can lead to a hurried and incorrect analysis or an untimely product. Large scale applications may take several months to be properly completed.

Finally the product of the system, a conditional projection of the impacts of activities and actions, is at best an educated guess and at worst a false prophecy of doom or cheer. Such products are difficult to vindicate historically since they will most often be wrong. But that is their purpose -- to alert us to situations that may lead to harm or to opportunities that may achieve beneficial results -- to stimulate change. In doing so, the conditions upon which the system has been developed and calibrated will also change as will the functional relationships that have been designed into it. If such endeavors are to be useful then perhaps historical vindication should not be such an important criteria for acceptance or rejection.

SEAS, when applied in the proper manner by experienced analysts, does aid in the tasks set forth in the introduction to this chapter. It can help to identify future constraints and societal choices, provide information useful to developing interim strategies consistent with long term social goals and, by providing explicit quantified conditional analysis, heighten our awareness of the tradeoffs inherent in the results our actions.

The papers that follow in this section of the book describe in detail the varied specific uses of SEAS.

BIBLIOGRAPHY

Almon, Clopper, M.R. Buckley, L.M. Horowitz and T. C. Reimbold, (1974). 1985: Interindustry Forecasts of the American Economy. Lexington, MA: D.C. Heath, (1974).

Forrester, Jay W. World Dynamics, Cambridge, MA: Wright-Allen Press, (1971).

Hoffman, K. C. and P.F. Palmedo Reference Energy Systems and Resources Data for Use in the Assessement of Energy Technologies, Rep. No. AET-8, Associated Universities Inc., Upton, New York, (1972).

House, Peter, Trading off Environment, Economics and Energy, Lexington, MA: D.C. Heath, (1977).

Lakshmanan, T.R. and S. Krishnamurthi, The SEAS Test Model: Design and Implementation, Washington, D.C.: U.S. Environmental Protection Agency, (1973).

Lakshmanan, T.R., P. Kroll, M. Pappas, L. Chatterjee, and B. Barron, The SEAS Region Model: An Assessment of Current Status and Prospects, Washington, D.C.: U.S. Environmental Protection Agency, (1979).

Lakshmanan, T. R. and Sam Ratick, 'Integrated Models for Economic-Energy-Environmental Impact Analysis,' Economic-Environmental-Energy Interactions, Edited by T.R. Lakshmanan and P. Nijkamp, Boston, MA: Martinus Nijoff Publishing, (1980).

Lakshmanan, T.R. and Sam Ratick, 'The Economic Environmental Effects of Energy Development Scenarios: A Strategic Assessment.' Paper presented at the Soviet-American Seminar on Urban Environments, Northwestern University, Evanston, Ill., (1977).

Placet, Marylynn and Donald Cooper, The Energy System Network Simulator, (ESNS) - Application Guide, Vol. 1, Prepared for the MITRE Corporation by International Research and Technology Corporation. Draft, (1979).

Ridker, Ronald G. and William D. Watson, (1980). To Choose a Future, Washington, D.C.: Resources for the Future, (1980).

U.S. Congress, Environmental Challenges of the President's Energy Plan: Implications for Research and Development, Washington, D.C., report prepared for The Subcommittee on the Environment and the Atmosphere of the Committee of Science and Technology, U.S. House of Representatives, Congressional Research Service, Library of Congress, GPO No. 052-010-04274-5, (1977).

U.S. Department of Energy, _National Environmental Impact Projection I_, Washington, D.C.: The MITRE Corporation, MTR-7905, (1979).

U.S. Department of Energy, _National Environmental Impact Projection II_, Washington, D.C. Final Report contract No. DE-AC03-79EV10092, Office of Environmental Assessment, (1980).

U.S. Department of Energy _Introduction to the Strategic Environmental Assessment System_, Washington, D.C.: The MITRE Corporation, MTR-80W115, (1980).

U.S. Department of Energy, _A Guide to the SEAS Documentation_, Washington, D.C.: The MITRE Corporation, MTR-19W00355, (1980).

11 The Economic and Environmental Consequences of Solar Energy Development

N. G. DOSSANI, WILLIAM D. WATSON AND W. P. WEYGANDT

1. INTRODUCTION

The purpose of this paper is to assess the economic and environmental consequences for the United States of using more solar energy in the next 25 years. What levels of investment are required to provide specific amounts of solar energy? What are the impacts on the size and composition of Gross National Product (GNP), on output by sector, and on pollution abatement expenditures? What are the likely impacts on the environment in different regions?

When it is possible and whenever it seems reasonable, our approach is to quantify impacts and compare scenarios having different amounts of solar energy. The Strategic Environmental Assessment System (SEAS) as currently maintained by the Department of Energy (DOE) is the major analytic tool used for making the assessments. Seven different solar scenarios, three developed by the Domestic Policy Review (DPR) panel on solar energy and four others developed by The MITRE Corporation under the DOE contract are analyzed. The analysis here is drawn from runs of SEAS undertaken for the Solar Energy Research Institute with the DPR scenarios. The DPR scenarios are of most interest since they were defined as part of a Federal assessment of solar energy and are, therefore, likely to be the basis for national solar policy. The MITRE scenarios differ from the DPR scenarios mainly in terms of level and mix of specific types of solar technologies and are used as "sensitivity tests" to see if economic and environmental impacts from the DPR scenarios change significantly as the level and mix of solar technologies change.

The next section provides a general description of SEAS and the DPR scenarios. Thereafter, we focus on impacts, first economic and then environmental impacts. These partial results are then put together; this provides measures of the total national value of solar energy for the DPR scenarios which are compared with similar measures for the MITRE scenarios. The final section is concerned with broad implications and qualifications.

Framework

The Domestic Policy Review of solar energy was announced by the President in his speech on Sunday, May 3, 1978. Its objectives were to assess the potential contributions of solar technologies to national energy supply by the year 2000, to estimate the amount of subsidy needed to achieve alternative levels of solar energy use, and to evaluate environmental, institutional, and economic impacts from using more solar energy.

Three alternative scenarios - base, maximum practical, and technical limits - were analyzed by the DPR (DPR 1978a, 1978b). Energy forecasts for each are shown in Table 1. The base or business-as-usual case takes into account the solar provisions of the National Energy Act and also assumes that other current Federal efforts to promote solar energy are continued. For the maximum practical scenario, the DPR estimated what might be achieved over the base case with a set of comprehensive and aggressive initiatives. In the technical limits scenario, growth in solar energy is constrained only by productive capacity. All three scenarios assume a growth in total energy demand from 72.6 quadrillion Btu (quads) in 1975 to 114 quads in 2000; a rise in the real price of petroleum to $25 per barrel by 2000 and a rise in GNP from $1.2 trillion in 1975 to $2.7 trillion by 2000 (in constant 1972 dollars).[1] In the DPR scenarios,each quad of solar energy displaces approximately 0.3 quads each of coal, oil and nuclear energy; and 0.1 quad of natural gas.

The system of models used in this analysis, SEAS, is a set of interlinked models, the core of which is the University of Maryland's dynamic input-output model of the United States economy, INFORUM (Almon, et al., 1974). Other components involve physical and monetary variables associated with energy, transportation, and the environment (residuals, abatement costs, and pollution damages) both at the national and regional levels (Table 2), (Figure 1).

The economic structure consists of 200 sectors delivering commodities to each other and to various consumers (households, investors in fixed capital and inventories, government and net exports). In addition, there are 294 side equations dealing with product and technology mixes within these sectors. The purpose of these side equations is to provide more detail for projecting energy use, residuals, and abatement costs. Abatement costs, together with energy investment costs, are calculated each year for over 300 sectors. These costs create a demand for resources, stimulating the output of sectors which directly or indirectly supply materials required to meet the specified levels of abatement and energy investment.

National gross pollution residuals for stationary sources are calculated by applying gross pollution coefficients (units of gross pollution per unit of output)[2] to output and side equation values. In any given year, more than 900 calculations are made. Net emissions are calculated as the product of gross emissions and the percent not controlled. These percentages correspond directly with the control levels and timing used in calculating abatement costs. National residuals are assigned to regions using employment shares from U.S. Water Resources Council (1974) and special industry location studies. In some cases, (for example, electric utility and industrial boilers)

---

[1] The currrent nominal price for petroleum of $28 per barrel is equivalent to a price of about $19 in 1972

[2] Output is measured either in dollars or physical units such as Btu's.

TABLE 1: DPR ENERGY FORECASTS[a]
(in quadrillion Btu)

	1975	2000		
		Base	Maximum Practical	Technical Limits
Coal	14.6 (20.)	38.8 (34.)	36.1 (31.)	33.8 (30.)
Oil	33.0 (45.)	32.4 (28.)	30.0 (26.)	26.8 (24.)
Gas	20.0 (28.)	18.0 (16.)	17.3 (15.)	15.9 (14.)
Nuclear	1.8 (2.)	15.0 (13.)	13.0 (11.)	9.0 (8.)
Geothermal	--	.5	.8 (1.)	1.0 (1.)
Solar	--	6.0 (5.)	12.9 (11.)	23.5 (21.)
Hydro	3.2 (4.)	3.3 (3.)	3.9 (3.)	4.0 (4.)
TOTAL	72.6	114.0	114.0	114.0
Imports[b]				
Oil	12.5	19.0	17.4	15.7
Gas	1.0	0.9	0.9	0.8

[a] For solar energy, estimates are quads of primary fuel displaced. Numbers in parentheses are percentages of totals. Numbers may not sum to totals because of rounding.

[b] Included in above totals.

TABLE 2. GROSS NATIONAL PRODUCT ACCOUNTS AND POPULATION[a] (billions of 1972 dollars and percentages)

	1975	1985	2000		
	Base	Base	Base	Maximum Practical	Technical Limits
Gross National Product	1,200	1,865	2,687	2,687	2,687
Consumption	768	1,192	1.785	1,771	1,754
Percent of GNP	64	64	66	66	65
Gross Private Domestic Investment	142	318	422	438	455
Percent of GNP	12	17	16	16	17
Government	266	368	481	480	478
Percent of GNP	22	20	18	18	18
Net Exports	24	-14	-1	-1	-3
Percent of GNP	2	---	---	---	---
Population	214	234	262	262	262

Note: Numbers may not sum to total due to rounding; --- represents zero or rounds to zero.

[a] Population in millions.

base case in order to release the resources needed for capital expenditures in the solar energy sectors.[5]

Included in the output of the modelling system is a projection of the total costs of pollution abatement, including capital and operating costs for water and air pollution abatement. As a share of total GNP these are estimated to range from 1.4 to 1.7 percent over the forecast period:

[5] This assumes that GNP is held constant. In the concluding section of this paper we analyze whether "potential" GNP will be higher as a result of the shifting composition of industrial jobs.

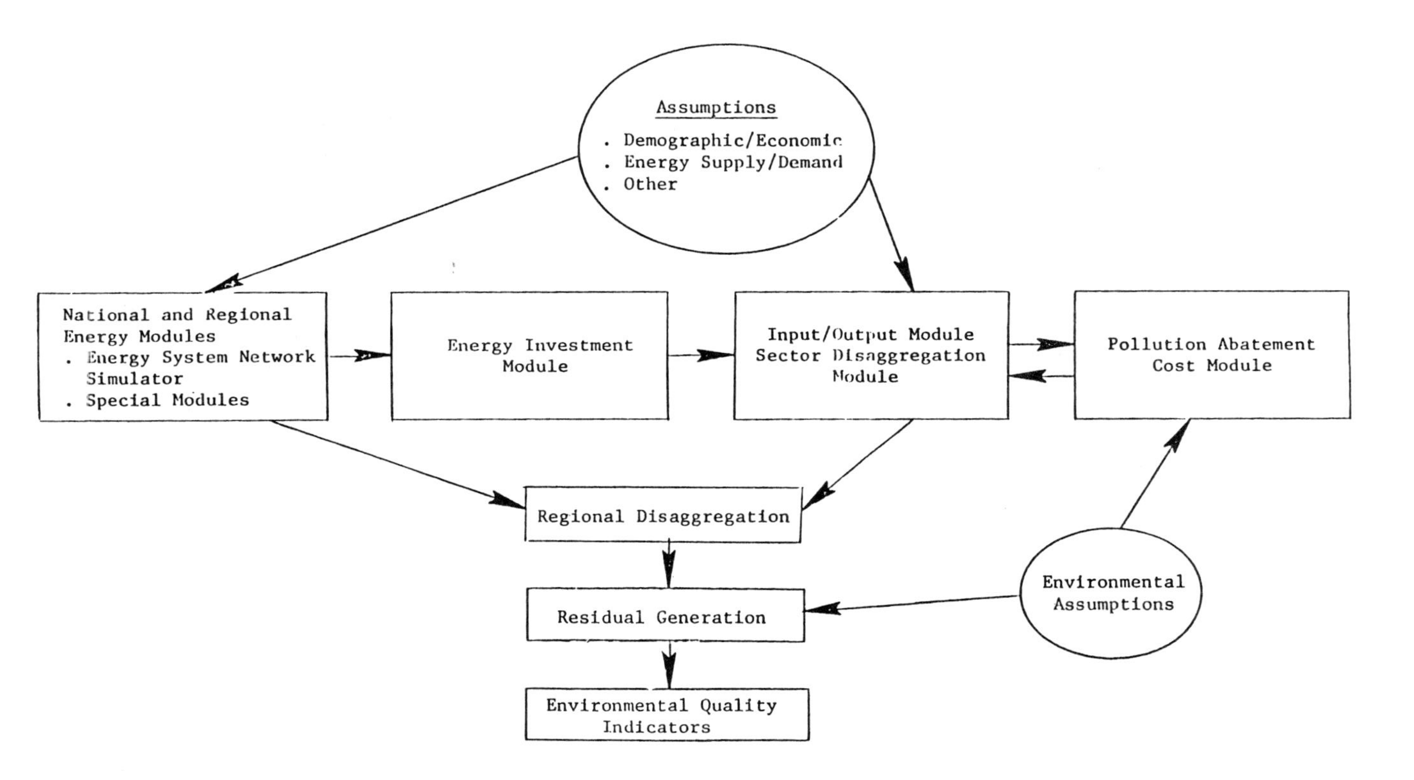

Figure 1. SEAS Block Diagram.

residuals are calculated initially at the regional level where region-specific characteristics can be included.

In addition to point source residuals, SEAS also calculates residuals for nonpoint sources: transportation, urban runoff, agriculture, mining, non-urban construction, and forestry. In all cases, estimates are made at the regional level.

A three-step procedure is used to estimate air and water pollution damages. Regional residuals are transformed into ambient concentrations using dispersion models with appropriate transfer coefficients for each region; per capita average damages in dollars are calculated as a function of average per capita exposure; and per capita damages are multiplied by regional population and summed to obtain national damages. Pollution damages are calculated only for the major air and water pollution residuals (particulates, sulfur oxides, nitrogen oxides, hydrocarbons, carbon monoxide, biochemical oxygen demand, chemical oxygen demand, suspended solids, dissolved solids, nutrients, oil and grease, bases, and acids).

The procedure for running the model involves interactions between the model and its users that are difficult to describe accurately in a brief space. In the first instance, the model is targeted on the GNP, energy forecasts and energy investment requirements in each of the DPR scenarios. Resource requirements for specific energy technologies are estimated within an energy investment sub-model that feeds back those requirements to the input-output model's transaction matrix. Household and state and local governments purchases are determined as a function of disposable income. These and other final demands determine output levels by sector, the latter determining non-energy investment requirements. If aggregate expenditures are greater (or less) than targeted GNP, disposable income is lowered (or raised) and the model rerun until equality with the GNP target is achieved. Since SEAS is an input-output model it tracks both direct (for example, steel for making solar collectors) and indirect (for example, coal for making the steel in solar collectors) impacts. The final set of estimated outputs is a consistent reflection of the GNP and energy targets in the DPR scenarios and the resource requirements (direct and indirect) for solar technologies.[3]

Economic Impacts

The total level of GNP is assumed in our analysis to grow at a rate of about 3.3 percent a year between 1975 and 2000[4] (or about 2.4 percent a year on a per capita basis). While the level of GNP is a scenario input, the composition of GNP is determined endogenously (Table 2). In particular the shift toward solar energy results in a significant decline in the levels of consumption when compared to the baseline. Consumption in the year 2000 in the technical limits scenario is projected to decline by $31 billion from its level in the

[3] A technical description of SEAS is provided in U.S. Department of Energy (1977). Details on the model that estimates pollution damages are provided in the Appendix to Ridker and Watson (1980).

[4] If a nonrecession year, 1977, is used as the base, the rate drops to 3.1 percent.

	1975	2000		
	Base	Base	Maximum Practical	Technical Limits
Pollution Abatement Costs (billions of 1972 dollars)	20.5	38.6	37.8	37.3
Percent of GNP	1.7	1.4	1.4	1.4

Increased use of solar energy is not expected to reduce pollution abatement expenditures significantly (e.g., for scrubbers and precipitators). The decline in pollution abatement costs for energy sectors is modest and is partially offset by an increase in pollution abatement costs for such non-energy sectors as fabricated metal products that supply direct and indirect inputs for solar investment.

Investment expenditures in energy facilities are shown in Table 3. In the base case, where coal, oil and nuclear fuel sources provide more than 75% of the domestic energy demand, total energy investment is forecast to grow at an annual average rate of 3.7 percent. These costs grow on average at 4.6 and 5.4 percent annually in the maximum practical and technical limits scenarios respectively, reflecting the greater capital intensiveness of solar and biomass technologies relative to conventional ones.

Comparison of energy investment costs among scenarios show the petroleum, synthetic fuels, and nuclear fuel cycles to register the largest declines in facility investment as solar energy forms a larger share of the U.S. fuel mix. For example, investment in the nuclear cycle of which more than 60 percent is for the construction of light water reactors,[6] is less in the maximum practical and technical limits DPR scenarios because more (both absolutely and relatively) solar and biomass, and less conventional fuels are being used to produce electricity.

It is estimated that between $23 and $45 billion of solar-related investment will be required to meet the solar energy demands of the DPR maximum practical and technical limits scenarios in the year 2000.[7] Thus, investment in solar facilities in 2000 alone could be nearly twice the total 1975 U.S. energy investment, and compares against a base case solar investment of about $66 billion for the entire 1975-2000 period. As Table 4 shows, most of the additional solar-related investment is for the construction of solar electric generating units. By 2000, investment in solar thermal, wind energy, and residential and centralized photovoltaic stations in the high solar scenarios will account for between 42 and 65 percent of allsolar-related facility investment. Most of the solar utility investment will be for centralized generating units. In all,

---

[6] The remaining 40 percent is for investment in uranium mining and conversion facilities.

[7] Over the forecast period (i.e., 1975-2000) solar investment amounts to $292 billion in the maximum practical scenario and $614 billion in the technical limits scenario.

TABLE 3. ENERGY FACILITY INVESTMENT COSTS BY FUEL CYCLE: COMPARISON OF BASE CASE TO MAXIMUM PRACTICAL AND TECHNICAL LIMITS SCENARIO (millions of 1972 dollars)

	1975	2000		
	Base	Base	Maximum Practical	Technical Limits
Total Energy Investment	25,071	62,285	76,717	93,460
Coal	5,845	11,723	10,245	9,883
Percent of Base			87	84
Synthetic Fuels	---	1,450	1,228	722
Percent of Base	---		85	50
Natural Gas	1,991	2,311	2,192	2,010
Percent of Base			95	87
Petroleum	7,179	6,906	5,826	3,781
Percent of Base			84	55
Nuclear Fuels	3,246	6,711	5,135	2,158
Percent of Base			77	32
Solar	---	5,695	23,423	45,294
Percent of Base	---		411	795
Geothermal	---	379	592	764
Percent of Base	---		156	202
Hydroelectric	525	175	591	694
Percent of Base			338	397
Electric Transmission/ Distribution	6,285	10,869	10,723	10,136
Percent of Base			99	93
Conservation	---	15,495	15,495	15,495
Percent of Base	---		100	100
Biomass	---	571	1,267	2,523
Percent of Base	---		222	442

Note: (---) Represents zero or rounds to zero.

TABLE 4: ENERGY INVESTMENT IN SOLAR FACILITIES IN 2000
(millions of 1972 dollars)

	Base	Maximum Practical	Technical Limits
Total Solar	5,695	23,421	45,294
Solar Thermal	401	3,577	6,105
Percent of Total	7	15	13
Residential Photovoltaic	301	4,244	11,819
Percent of Total	5	18	26
Centralized Photovoltaic	397	3,468	9,543
Percent of Total	7	15	21
Solar Heating and Cooling of Buildings	1,024	3,328	5,484
Percent of Total	18	14	12
Agriculture Industrial Process Heat	2,751	4,521	7,722
Percent of Total	48	19	17
Wind Energy Conversion System	375	962	1,885
Percent of Total	7	4	4
Passive Space Heating	446	3,321	2,736
Percent of Total	8	14	6

Note: Numbers may not sum to totals due to rounding.

approximately 35 to 40 percent of all solar facility investment in 2000 in both the maximum practical and technical limits scenarios will be for centralized solar systems, whereas only 21% of solar investment in the base case is for centralized solar facilities. In summary, not only is there projected to be a significant increase in solar-related energy investment from the base case to the maximum practical and technical limits scenarios, but there is a noticeable shift in the centralized vs. decentralized or soft vs. hard composition of the investment.

National and Regional Emissions

For the base case at the national level, it is estimated that discharges for most of the major air and water pollutants will decline over time (Table 5). This reflects implementation of Federal and state pollution control regulations. Discharges of nitrogen oxides, an exception, increase because of the growing amount of coal burned in power plants where abatement control is not sufficient to offset the increase in gross tonnage. Other emissions, such as solid wastes, are estimated to increase substantially over time as population and the

TABLE 5: NATIONAL EMISSIONS[a]

(million tons)	1975 (million tons)	1985[a] (million tons)	1990 Base (million tons)	1990 Maximum Practical (Percent of Base)	1990 Technical Limits (Percent of Base)	2000 Base (million tons)	2000 Maximum Practical (Percent of Base)	2000 Technical Limits (Percent of Base)
Particulate Matter	13.8	8.0	8.7	101	101	10.0	104	108
Sulfur Oxides	29.6	26.8	27.8	100	99	26.3	94	88
Nitrogen Oxides	19.1	21.5	21.6	99	99	23.3	94	91
Hydrocarbons	14.2	10.5	9.7	100	100	10.0	99	97
Carbon Monoxide	99.9	75.0	58.3	100	100	59.9	100	100
Biochemical Oxygen Demand	6.3	3.9	4.1	100	100	4.0	100	100
Suspended Solids	496.2	348.3	366.2	100	100	336.9	100	100
Dissolved Solids	260.0	162.3	173.6	100	100	166.5	99	97
Solid Wastes	67.8	116.2	141.1	100	99	170.8	97	96
Industrial Sludges	7.8	30.8	59.8	98	96	107.7	90	89

[a] Includes point and nonpoint source residuals.

[b] Emissions are the same in all three scenarios.

economy grow. Captured or controlled air and water emissions and scrubbing materials become waste discharges to land, "industrial sludges" in our system, and these are also projected to increase by substantial amounts.

The use of more solar energy leads to both increases and decreases in residuals. On the one hand, solar energy generates its own brand of pollution including particulates from wood burning, biochemical oxygen demand and suspended solids from biomass farms, and pollution from the manufacture of solar equipment (solar indirect residuals). On the other hand, solar energy reduces some residuals. For example, it displaces other energy forms like coal and oil where direct residuals are relatively high. Also, as the manufacture of solar equipment increases (which, as noted, raises solar _indirect_ residuals) the manufacture of other things declines leading to an offsetting influence on _indirect_ residuals.

Solar energy, with a few exceptions, generates fewer residuals overall. One exception, an increase in particulate matter, occurs because particulates from combustion of wood and other biomass energy systems are almost completely uncontrolled. For other pollutants, discharges are lower in the higher solar scenarios. In the year 2000 as compared to the base case, the technical limits scenario has 8% more particulates, 12% fewer sulfur oxides, 9% fewer nitrogen oxides, 3% less dissolved solids and 11% less industrial sludges.

At the Federal region level, residuals in the higher solar scenarios are also generally lower. However, some regions are impacted to a greater or lesser extent than the nation depending upon their economic base. For example, while sulfur oxides for the nation are 12% lower in 2000 under the technical limits scenario, the New England and New York/New Jersey Federal regions have 34% and 24% fewer sulfur oxides, repectively (Table 6). There, reductions result from a decline in coal and oil combustion and small net increases in indirect sulfur oxide discharges.

It should be noted that the estimates in Table 6 for biochemical oxygen demand and suspended solids cover only discharges from point sources and solar technologies. Residuals for nonpoint sources are not included but had they been, the differences for Biochemical Oxygen Demand would be closer to zero and negative in some regions reflecting reductions in emissions from reduced levels of mining (mainly coal) and forestry as more solar is added. But even with this adjustment, the Central Federal region would still show a relatively large percentage increase for suspended solids in the higher solar scenarios owing to the solar biomass farms that are located there because of favorable economics.

National and Regional Pollution Damages

A hierarchy of measured environmental impacts would begin with tons of residuals followed in order by ambient concentrations, exposures to values at risk, physical impacts on damaged objects including health impacts to individuals, and the translation of physical impacts into dollar damages. This is the sequence followed by the model we use for estimating air and water pollution damages. The most encompassing

TABLE 6: REGIONAL EMISSION TRENDS IN 2000
(Percent)

	Particulate Matter	Sulfur Oxide	Biochemical Demand[a]	Suspended Solids[a]	Industrial Sludges
A. Maximum Practical vs. Base Case					
New England	20	+2	---	4	---
New York/New Jersey	4	-19	-0.2	2	35
Middle Atlantic	3	-5	-0.2	3	-9
Southeast	3	-5	-0.3	12	-5
Great Lakes	4	-7	-0.2	14	-17
South Central	-	-11	-0.7	16	-28
Central	2	-1	-0.1	68	3
Mountain	7	-7	-0.1	17	-5
West	2	-6	-0.1	1	24
Northwest	12	-2	-0.4	15	19
B. Technical Limits vs. Base Case					
New England	37	-34	-0.4	12	162
New York/New Jersey	9	-24	-0.4	8	111
Middle Atlantic	5	-12	-0.4	11	-23
Southeast	5	-10	-0.9	44	-9
Great Lakes	10	-9	-0.5	49	-8
South Central	3	-14	-1.3	56	-36
Central	2	-10	-0.7	234	-31
Mountain	13	-16	-0.5	63	-11
West	1	-14	-0.3	5	12
Northwest	22	-11	-0.9	53	35

(---)indicates zero or rounds to zero.

[a] Discharges from point sources and solar technologies only.

measure of environmental impacts for our purposes are the estimates of dollar damages.

As indicated in Table 7, national pollution damages increase over time in all scenarios with smaller increases occurring in the higher solar scenarios. These increases occur, in spite of generally lower residual discharges over time, because population and values at risk are increasing over time, a factor which is captured in the estimates. Nonetheless, solar energy has a significant impact on pollution damages. For example, by 2000 the technical limits and maximum practical scenarios have $23 and $16 billion fewer damages than the base case, respectively.

TABLE 7: NATIONAL POLLUTION DAMAGES
(billions of 1972 dollars)[a]

	1975	1985	1990 Base	1990 Maximum Practical	1990 Technical Limits	2000 Base	2000 Maximum Practical	2000 Technical Limits
Air	76.6	104.2	113.7	112.6	109.8	109.8	96.4	91.4
Water	25.7	27.4	32.3	32.0	31.5	38.3	36.2	34.2
Total	102.3	131.6	146.0	144.6	141.3	148.1	132.6	125.6

[a] For the purpose of evaluating adverse health impacts, specifically mortality, the statistical value of life is assumed to be $200,000 in 1975 (Thaler and Rosen, 1976) rising to $434,000 by 2000. Costs of illness are assumed to be $20,000 per illness in 1975 and rising to $43,000 by 2000. The increase over time reflects increased willingness to pay for a cleaner and safer environment as income rises (Ridker and Watson, 1980, Appendix).

While of interest as a summary measure, national damage estimates are not very illuminating. National estimates are the sum of estimates made for individual air and water sheds; it is the regional estimates that reveal more clearly the difference that solar energy makes.

Air pollution damages are estimated for 243 air quality control regions (AQCR's). A comparison of the maximum practical over the base scenario indicates that in the year 2000, air pollution damages decrease in 171 AQCR's and increase in 72 AQCR's. The regions experiencing the largest changes are shown in Table 8.

The AQCR's that have fewer air pollution damages display several different patterns. In some AQCR's (Los Angeles, for example) damages from particulates are lower because petroleum refining and fuels combustion is reduced by the use of solar energy. In other AQCR's (New York City/New Jersey, Chicago, and Cleveland, for example) particulate damage increases because the reduction in damages from less fuel combustion does not offset the increase from biomass combustion and activity in the stone and clay products sector (the latter being an indirect impact from increase solar energy use).[8] In contrast, for these same AQCR's, damages from sulfur and nitrogen oxides are reduced because solar energy displaces a significant amount of fuel combustion in electric power plants and industrial and commercial boilers. And overall total regional damages fall because these damage reductions are larger than the damage increases from particulates.

For AQCR's with larger damages, what happens in some cases (for example, Buffalo) is that the increase in damage from biomass combustion and stone and clay production is not offset by the reduced damages from less conventional fuel combustion. In other cases (Central Michigan and Spokane, for example) the increase in stone and clay production requires the installation of coal-fired electric power plants and their discharges account for the increase in damages.

Water pollution damages are estimated for 101 aggregated sub-areas (ASA's). Compared to damages in the base case, 81 ASA's have lower damages in the maximum practical scenario in 2000. For ASA's that experience increases, the change is relatively small, with the largest increase ($71 million in 2000 in the maximum practical scenario) occurring in San Francisco Bay, (Table 9). The decline in damages in most regions is accounted for by fewer damages from reduced levels of electric generation and coal mining. Regions that experience large increases in damages have higher levels of residuals from the increased use of solar energy in their industrial sectors and from the increased output of electric energy which is needed to support regional production of solar energy equipment.

National Value of Solar Energy

A useful way to summarize the results we have obtained is to indicate those scenarios having favorable economic/environment trade-offs. Ideally, we should translate all the changes observed over time and between scenarios into a single common denominator of welfare. While this is impossible given the diversity of environmental and economic impacts from solar energy, it is possible to estimate some, hopefully the major, costs and benefits that occur when solar energy increases.

Four categories of costs and benefits have been quantified. The first category is change in total national consumption including consumption of state and local government services. This occurs as a result of the following shifts in the composition of GNP as use of solar energy is increased:

(a) Solar energy is more costly than conventional energy. For a given GNP (near full employment), this means that the solar investment component of GNP rises as solar use increases and other components decline.[9] In our economic model, the Federal government component, exports and a large part of non-energy investment are held constant and consequently the adjustment occurs

---

[8] This is mainly output of cement for building supports for solar equipment or structures where solar equipment is manufactured.

TABLE 8: AIR POLLUTION DAMAGES IN 2000 BY REGION, MAXIMUM PRACTICAL VS. BASE
(millions of 1972 dollars)

Regions with Fewer Damages[a]

Region	AQCR[c]	Change in Damages
NYC-NJ	43	$8,700
Los Angeles	24	1,620
Chicago	67	1,230
Boston	119	340
Washington,D.C.	47	190
Cleveland	174	140
St. Louis	70	140
Dallas	215	120
Houston	216	107
Salt Lake City	220	89
Dayton	173	86
Hartford	42	71
Denver	36	70
Mobile	5	70
Hampton Roads	223	64
Seattle	229	63
Portland	193	60
Atlanta	56	59
Baltimore	115	58
Philadelphia	45	52
New Orleans	106	52
Indianapolis	80	52
Jacksonville	49	38
Birmingham	4	34
Columbus	176	32
Cincinnati	79	29
Charlotte	167	28
Minneapolis	131	28

Regions with More Damages[b]

Region	AQCR	Change in Damages
Detroit	123	$161
Bakersfield	31	69
Sacremento Valley	28	49
Buffalo	162	44
Kansas City	94	28
San Francisco	30	26
Milwaukee	239	23
Allentown	151	20
Spokane	62	19
Central NJ	150	14
Central Michigan	122	13
Memphis	18	12

[a] All regions where damages decrease by at least $25 million.

[b] All regions where damages increase by at least $10 million.

[c] AQCR - Air Quality Control Region.

TABLE 9: WATER POLLUTION DAMAGES IN 2000 BY REGION, MAXIMUM PRACTICAL vs. BASE (millions of 1972 dollars)

Regions With Fewer Damages[a]

Region	ASA[b]	Changes in Damages
Texas Gulf - Trinty R.	1,202	$614
Mid-Ohio River Basin	502	204
Southern Florida	305	195
NYC-NJ	202	157
Lake Erie West	406	137
South Coast - California	1,806	107
Northwest Utah	1,604	105
Ohio River Basin - Ohio	503	64
Ohio-Kanawha R.B.	504	47
Lower Arkansas R.B.	1,104	46
Northern Florida	304	43
Northeast California	1,802	40
Central Texas	1,204	33
Southern Arizona	1,503	32
Tombigbee, R.B.	308	27
Sabine - Neches R.B.	1,201	26
Mississippi R.B. - Missouri	705	25
Ohio - White - Patako R.B.	506	25

Regions With More Damages[c]

Region	ASA	Changes in Damages
San Francisco	1,804	$71
San Joaquin R.B.	1,803	48
Lake Ontario	408	20
Delaware R.B.	203	13

[a] All regions where damages decrease by at least $25 million.

[b] ASA - Aggregated Sub-area.

[c] All regions where damages increase by at least $10 million.

by lowering the rate of growth of national consumption and state and local government services.[10]

(b) Solar energy displaces petroleum and natural gas and consequently smaller amounts of these high cost fuels are imported. The resources used to pay the import bill can alternatively be used to buy other imports, more investment, or more domestic consumption. The reduced cost of imported petroleum and natural gas is a net economic gain from solar energy.

(c) Increased use of solar energy allows more total output from a given labor force. In the high solar scenarios, it turns out that workers are shifted into sectors where their productivities are higher. Thus, a given labor force is able to produce more output. About 35 percent of this is available for consumption and it offsets some of the consumption losses directly associated with higher solar investment costs.[11]

The second category of change associated with increased use of solar energy is reduction in occupational accidents. The number of occupational deaths are slightly higher in the solar cases but occupational injuries are less. For evaluation in dollars, costs of $200,000 per death and $20,000 per injury rising to $434,000 and $43,400, respectively, by 2000 are used (see footnote to Table 7). At these costs, the higher solar scenarios show savings when compared to the base case.

---

9 Alternatively, we could have allowed some of the extra solar cost to displace normal productive investment -- a situation more in line with what might actually happen. In this case, capital-labor ratios would fall and GNP would grow less rapidly leading eventually to even lower consumption levels than in our simulations. Thus, our estimated reductions in consumption are likely to be conservative. See Ridker, _et al_. (1977) for an extended discussion of this point.

10 Extra investment costs for solar energy would result in a slowdown in the _rate of growth_ of economic welfare measured in conventional terms, _not a reduction_ in absolute levels. For example, for the period from 1980 to 2000, aggregate national consumption is estimated to increase at these annual rates (scenario in parentheses): 3.1% (base), 3.0% (maximum practical); and 2.9% (technical limits).

11 This shift was not incorporated into the estimates of residuals and pollution damage reported earlier. It would increase the amount of residuals and pollution damages above the forecast levels. Thus, the correct measure of gain in this case is the increase in consumption minus any increase in associated environmental damages.The increase in consumption reported in Exhibit 11 below nets out the associated increase in environmental damages.

TABLE 10: NATIONAL VALUE OF SOLAR ENERGY, DPR SCENARIOS

	Maximum Practical	Technical Limits
A. Change over Base Case (cumulative from 1975 to 2000)		
Total Private Consumption and S&I Govt. Services	-$101	-$290
Energy Investment Costs	(-188)	(-454)
Output from Increased Productivity	(47)	(77)
Savings in Petroleum and Natural Gas Imports	(40)	(87)
Savings in Occupational Injury Costs	9	19
Savings in Nuclear Power Plant Health Costs	2	6
Savings in Air and Water Pollution Damages	98	164
	8	-101
B. Annual Net Benefits of Solar over Base Case beginning in 2001[a]	21	29
C. Year by Which Solar Installed to 2000 Provides New Positive Payoff over Base Case	2000	2004
D. Year for Positive Payoff After Adjustment for Conventional Back-up and Nuclear Phaseout Costs	2005	2012

[a] Cost savings in fuel, pollution damages, occupational accidents, and radioactive emissions minus operation and maintenance costs for solar systems.

A third category of change is reduction in health impacts from the radioactive emissions of nuclear power plants. The higher solar scenarios have less nuclear energy. This reduction in output results in fewer expected deaths.[12] The final category of change associated with increased use of solar energy is reduction in pollution damages. As shown in the previous section, pollution damages are less in the higher solar scenarios. This is a real economic gain for solar energy.

Gains and losses for the maximum practical and technical limits scenarios over the base case are shown in panel A of Table 10. Overall, the maximum practical scenario has $8 billion more cumulative economic welfare than the base case and the technical limits scenario has $101 billion less cumulative economic welfare.

However, the estimates in panel A are partial results. Solar energy has larger "front end" investment costs but could provide annual net savings over conventional energy when operating, maintenance and life-cycle environmental costs are compared. To see the implications of this, it is assumed that solar energy investment stops in 2000 and thereafter the solar capacity installed to 2000 is operated until it wears out and is replaced.

Estimates of net annual benefits starting in 2001 are provided in panel B of Table 10. Benefits become larger as solar energy increases. These annual benefits are divided into front end net costs to determine the year when solar energy begins to provide a positive net payoff. For the maximum practical case this is 2000; for the technical limits scenario a positive payoff occurs in 2004 (panel C).

But several import costs have been excluded and, consequently these results are biased toward making solar appear more favorable than it is likely to be. One cost is the expense of conventional energy backup i.e., the cost for conventional capacity that must be in place to supply energy when weather conditions are very unfavorable for solar energy. These costs are difficult to estimate and, among other factors, depend upon cost tradeoffs between solar energy storage and conventional peaking capacity. A very rough, back-of-the-envelope estimate of cumulative back-up costs for the maximum practical scenario (over the base case) is $100 billion rising to $135 billion in the technical limits scenario.

Another cost, applicable only in the technical limits scenario, is the extra cost associated with substantial reduction in nuclear energy. The DPR technical limits scenario imposes a nuclear phase-out. Nuclear power plants currently operating and under construction are allowed to operate for a normal life time but construction of any additional plants is halted. The foregone nuclear

---

[12] The Nuclear Energy Policy Study Group (NEPSG, 1977) estimated that .4 to 10.7 deaths per plant per year (1000 MW at a 70% capacity factor) can be expected from the operating and accidental radioactive emissions of nuclear power plants. The average of six deaths per plant year is used in the calculations. A summary of the NEPSG estimates is provided in Watson (1978).

capacity, mainly in the Northeast and Southeast United States and California, is displaced by solar energy and coal-fired capacity. Our estimates do include costs for the solar and coal-fired capacity that fills the gap but do not include costs for transporting coal over the longer distance required by a nuclear phase-out. These costs are estimated to be $90 billion in the period to 2000 (see Watson, 1978).

As shown in panel D of Table 10, adding conventional back-up and nuclear phase-out costs extends the payback period. It becomes 2005 for the maximum practical scenario and 2012 for the technical limits scenario.

The economic/environment trade-offs of the DPR maximum practical scenario relative to the base case are favorable. Even though its front-end costs could be as much as $100 billion higher than the base case, this could be paid off in a four to six year period. After that it would provide more economic welfare than the DPR base case. In contrast, the trade-offs in the DPR technical limits scenario appear to be unfavorable. At a zero discount rate, it would not show a positive net payoff relative to the base case until about 2012. At 6 percent, discounting net benefits back to the year 2000, a positive net payoff relative to the base case would not occur until the year 2020. More importantly, as we show below, there are other mixes of solar and conventional energy technologies which could provide the same level of solar energy as in the technical limits scenario but with shorter payback periods.

It is of interest to compare the DPR scenarios with some alternatives. Under contract to DOE The MITRE Corporation has developed some scenarios that have the same level of total energy as the DPR scenarios and are generally consistent with the macroeconomy in the DPR scenarios.[13] The MITRE scenarios have solar energy levels that cover about the same total range as the DPR scenarios but in composition, they differ considerably, having less photovoltaic solar energy, more solar thermal and passive energy, and more wind energy (Figure 2). Also the MITRE scenarios have solar displacing more coal and less nuclear energy so that in the higher solar cases there is no nuclear phase-out as in the DPR technical limits scenario.[14]

To compare the DPR and MITRE scenarios we have plotted (Figure 2). each scenario's cumulative solar energy against its payback period (at

---

[13] "Technology Assessment of Solar Energy: Description of Solar Technologies and Energy Scenarios," Dept. W-51, McLean, Virginia, Draft Working Paper 79 W00228, May 24, 1979.

[14] Benefits of solar energy for the MITRE scenarios were estimated by interpolation on the DPR estimates using solar energy use as a basis. An alternative would have beenm to do new runs in our economic model using the MITRE forecasts. However, it was judged that this was not necessary because data on residuals, total energy, and GNP would have been nearly the same. Furthermore, the DPR and MITRE projections of solar energy fall within a narrow band. Because of these considerations, linear interpolation using solar energy as a basis is probably a good first approximation to an independent forecast.

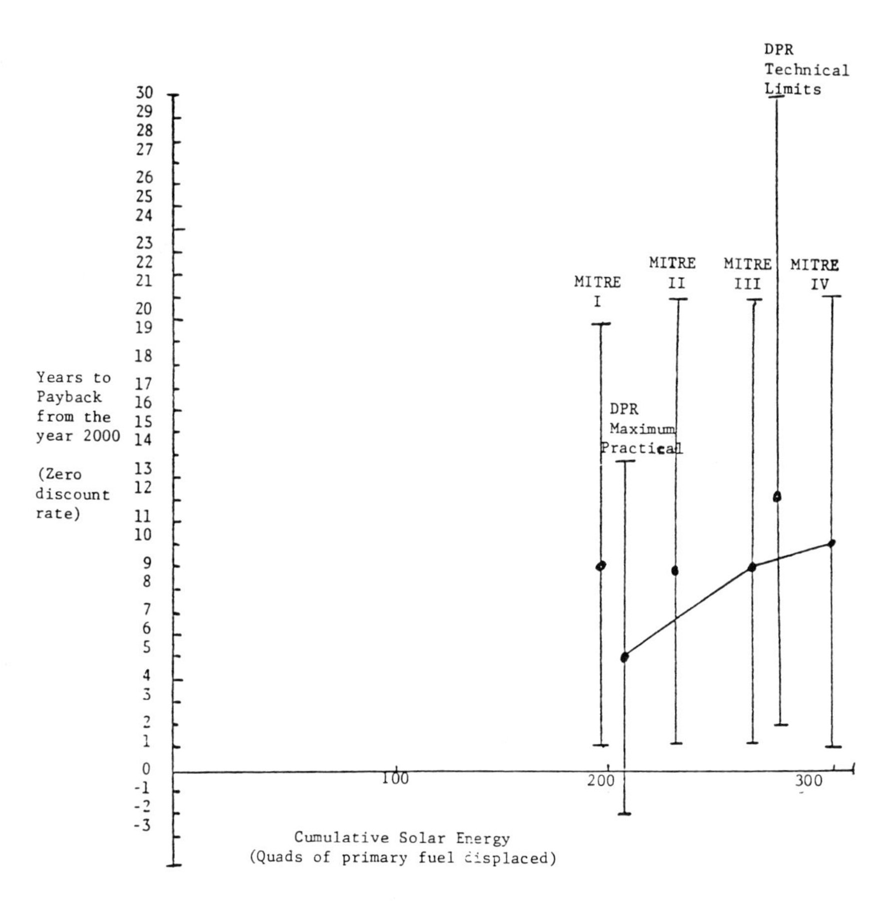

[a] Error bands are generated by using high and low values for estimating the savings in air and water pollution damages. The low value for the saving in air and water pollution damages is .4 times the mean or expected value; the high value is 2.3 times the mean or expected value. The error band is meant to cover the true value at a .95 probability level. Ridker and Watson (1980, Appendix) is the source of the low and high multipliers.

Figure 2. Comparison of DPR and MITRE Scenarios[a]

zero discount rate). To reflect uncertainty in our estimates ofpollution damages we have also calculated for each scenario a range of paybacks meant to cover the true value of pollution damages with a probability of .95. The payback ranges are shown as bars in Figure 2. The "dots" are expected payback periods. Economically, the most favorable places on the plot are at high solar energy levels and low payback periods since at these places the extra front end costs of solar energy are repaid by extra life-cycle benefits in the shortest time possible. As shown, from among the alternatives considered, the locus of favorable cases (using expected values) includes the DPR maximum practical scenario and the MITRE III and IV scenarios. MITRE III is clearly better than MITRE I and II, and MITRE IV is clearly better than the DPR technical limits scenario.

It is not surprising that the locus of expected values for the favorable cases has a positive slope. Less expensive solar technologies are exploited first; then, as the use of solar energy increases, costs increase as more use is made of higher cost solar technologies. On the benefits side, the initial amounts of solar energy save relatively large amounts of pollution damages. Then as the use of solar energy increases, incremental benefits, while positive, are less than preceding amounts since savings in pollution damages fall as pollution is reduced by solar energy. This combination of solar front end costs rising faster than benefits leads to the longer payback periods at the higher solar levels.

Conclusions and Qualifications

Our conclusions with respect to the economic and environmental consequences of solar energy depend upon which of four different classes of issues is considered. The first concerns impacts aggregated to the national level between now and the year 2000. In the period to 2000 increased use of solar energy leads to higher overall net costs. The substantial build-up in solar capacity generates high front end costs that exceed environmental and economic benefits. By the year 2000, compared to our base case, total net costs of higher solar energy levels could accummulate to several hundred billion dollars.

The second set of issues involves trade-offs between benefits in the period after 2000 and the earlier front end costs. To increase solar to higher levels it is necessary to turn increasingly to higher cost solar technologies (for example, photovoltaics and solar thermal electric systems). Environmental benefits, on the other hand, increase at declining rates as solar increases. Thus, as we have shown, the payback period is extended at higher solar energy levels. From the perspective of the year 2000, if society is willing to invest in a project that is expected to pay itself off in 10 years or less, then (discounting benefits at 6%) the DPR maximum practical scenario, according to our estimates, would be an acceptable project. The two other favorable scenarios that we have identified would fail this test: at 6% MITRE III has a payback period of 13 years with MITRE IV at 14 years.

Our finding that the DPR maximum practical scenario is economically better than a "free market" case (this being the DPR base case or even less solar energy) is consistent with economic theory. Solar energy generates substantial external environmental benefits that will

TABLE 11: SOLAR ENERGY, 2000
(Quads of Primary Fuel Displaced)

	DPR			MITRE			
	Base	Maximum Practical	Technical Limits	I	II	III	IV
Thermal	2.0	4.6	8.6	4.3	5.8	7.1	9.3
Passive	0.2	1.0	1.0	1.1	1.4	1.6	1.9
Photovoltaic	0.1	1.0	2.5	0.07	0.4	0.51	0.64
Wind	0.6	1.5	3.0	1.55	2.4	3.25	4.0
Biomass	2.9	. 4.2	6.5	2.83	3.89	4.82	5.38
STES[a]	0	0	0	0.01	0.05	0.1	0.2
OTES[b]	0	0.1	1.0	0	0.2	0.5	0.7
Hydro	3.3	3.9	4.0	3.4	3.4	3.4	3.4
Animal Waste	0.2	0.3	0.5	0.2	9.2	0.3	0.3
Other	0	0.2	0.4	0	0	0	0
Total	9.3	16.8	27.5	13.5	17.7	21.6	25.8
Cumulative 1975 to 2000	159	205	276	196	229	265	298

a Solar Thermal Electric System.

b Ocean Thermal Energy System

not be properly reflected in private decisions because individuals cannot capture all of these benefits. Therefore, the private market left alone will provide too little solar energy. Higher solar levels are achieved in the DPR maximum practical scenario because the federal government subsidizes solar energy and thus provides incentives to go beyond "free market" levels.[15] But as we have attempted to show this is quite appropriate because society from the perspective of the year 2000 could reap substantial net benefits -- on the order of $175 billion -- if the DPR maximum practical scenario were implemented.[16]

The third set of issues involves regional economic and environmental impacts and economic impacts on specific sectors. All regions and sectors will not necessarily benefit from increased use of solar energy. In the period to 2000, for example, environmental damages are estimated to increase in some regions as solar energy increases. It is unrealistic to ignore sector and regional impacts on the grounds that special programs could be designed to offset adverse impacts. Considerations such as these often block action and consequently this limitation must be kept in mind when evaluating the principal results presented in this study.

A final set of issues involves things that have been left out of the analysis or use of certain assumptions which if incorrect could lead to much different implications. A number of environmental impacts are not evaluated. This includes disturbed land, environmental risk from industrial sludges and highly toxic nuclear wastes, and impacts on climate from carbon dioxide build-up. Disturbed land could be higher with increased solar energy from biomass farms and windmills. In contrast, solar energy has an advantage in terms of reducing environmental risk from industrial sludges, nuclear wastes and carbon dioxide accumulation. Finally, the higher capital costs of the higher solar scenarios require a higher savings rate or a smaller amount of resources going into normal productive investments. If the extra investment for solar energy occurs at the expense of other normal productive investment, then economic growth could be damped and the macroeconomic costs of the higher solar scenarios could be understated by this study.

Thus we must end with a note of caution. When the above factors are more carefully assessed different consequences could result. But for the time being, our best estimate is that the nation would benefit by having solar energy at the levels and by the technologies indicated in the DPR maximum practical scenario.

---

[15] Assuming that the base case is a "free market" case, our estimates indicate that, in the period from 1980 to 2000 cumulative Federal subsidies of about $226 billion (1972 dollars) would have to be provided to increase solar energy from the levels in the base case to those indicated for the DPR maximum practical scenario.

[16] Calculated as net benefits of $21 billion discounted at 6% for 25 years minus front end costs of $92 billion. Similar calculations for the DPR technical limits and MITRE scenarios show that the DPR maximum practical scenario has the largest net benefit.

It is not surprising that the locus of expected values for the favorable cases has a positive slope. Less expensive solar technologies are exploited first; then, as the use of solar energy increases, costs increase as more use is made of higher cost solar technologies. On the benefits side, the initial amounts of solar energy save relatively large amounts of pollution damages. Then as the use of solar energy increases, incremental benefits, while positive, are less than preceding amounts since savings in pollution damages fall as pollution is reduced by solar energy. This combination of solar front end costs rising faster than benefits leads to the longer payback periods at the higher solar levels.

Conclusions and Qualifications

Our conclusions with respect to the economic and environmental consequences of solar energy depend upon which of four different classes of issues is considered. The first concerns impacts aggregated to the national level between now and the year 2000. In the period to 2000 increased use of solar energy leads to higher overall net costs. The substantial build-up in solar capacity generates high front end costs that exceed environmental and economic benefits. By the year 2000, compared to our base case, total net costs of higher solar energy levels could accummulate to several hundred billion dollars.

BIBLIOGRAPHY

Almon, C., Jr., M. B. Buckler, L.M. Horowitz, and T. C. Reimbold, 1985: Interindustry Forecasts of the American Economy. Lexington Books, Lexington, Massachusetts, (1974).

DPR (Domestic Policy Review), (1978a). Status Report on Solar Energy.August, Washington, D.C., (1978a).

DPR (Domestic Policy Review), Solar DPR Response Memorandum. November, Washington, D.C., (1978b)

NEPSG (Nuclear Energy Policy Study Group), Nuclear Power Issues and Choices. Vallinger Publishing Company, Cambridge, Massachusetts, (1977).

Ridker, R.G. and W.D. Watson, To Choose A Future: Resource and Environmental Consequences of Alternative Growth Paths. Johns Hopkins University Press for Resources for the Future, Baltimore, Maryland, (1980).

Ridker, R.G., W.D. Watson, and A. Shapanka, 'Economic,Energy and Environmental Consequences of Alternative Energy Regimes: An Application of the RFF/SEAS Modeling System,' in Modeling Energy-Economy Interactions: Five Approaches. Ed. by C. Hitch, Resources For the Future, Baltimore, Maryland, (1977).

Thaler, R. and S. Rosen, 'The Value of Saving a Life:Evidence from the Labor Market' in Household Production and Consumption. Ed. by Nelson Terleckyj, National Bureau of Economic Research, Washington, D.C., Vol. 40, pp. 265-298, (1976).

U.S. Department of Energy, User's Manual for the Strategic Environmental Assessment System (SEAS). Washington, D.C, (1977).

U.S. Water Resources Council, 1972 OBERSProjections,Regional Economic Activity in the U.S. USDA Economic Research Service, Washington, D.C., (1974)

Watson, W.D., 'Economic and Environmental Consequencesof a Nuclear Power Plant Phaseout' in The Journal of Energy and Development, Vol. III, April, No. 2, pp. 277-317, (1978).

12 An Energy Impact Assessment Modelling Technique for the Transportation Sector

PHILIP D. PATTERSON

1. INTRODUCTION

The transportation sector in the U.S. will use 20.8 Quads (9.8 MBPD) in 1980 and this will represent about 26 percent of all of the U.S. energy use. But the transportation sector will account for in excess of 53 percent of the U.S. petroleum use since it relies almost exclusively on petroleum products for its energy (the exceptions being natural gas for some pipelines and electricity for some pipelines and for some mass transit).

Therefore, measures to reduce transportation energy use or to switch to non-petroleum fuels are very important to this country's goal of becoming less dependent on imported oil. The Office of Transportation Programs (OTP) in the Department of Energy has as its objective the reduction of petroleum energy use in the transportation sector. OTP employs technological and policy measures in its effort to achieve this objective.

The major technological programs currently supported by OTP are those for gas turbine vehicles (autos, trucks, buses, and marine vessels), Stirling engine automobiles, electric and hybrid-electric light duty vehicles, and advances by heavy duty truck diesel engines.

The major non-technological programs of OTP are those dealing with training drivers to drive in a fuel efficient manner, distributing the gas mileage guide, promoting vanpools, and setting state gasoline consumption targets.

This paper will deal with one type of impact assessment technique that OTP has applied to its major technological programs. Of particular concern is the impact of direct energy savings in the transportation sector on indirect energy use related to this transportation conserving technology.

2. DIRECT AND INDIRECT ENERGY USE

Direct energy use in the transportation sector is limited to the fuel used in moving passengers and freight. For example, the direct energy use associated with an automobile is the gasoline or diesel fuel used to propel the vehicle. All other energy associated with the manufacture and use of the vehicle is the indirect energy. With respect to the automobile the indirect energy includes the power requirements of the auto factory (heat, power equipment, lights, etc.), the energy embodied in the materials used in the vehicle construction (mainly steel, glass, aluminum, rubber and plastics), and the energy embodied in the auto factory and its equipment. The indirect energy associated with the use of the automobile includes also such things as energy associated with auto insurance companies use, auto sales and repair facilities, highway construction and maintenance, parking garages, and gasoline service stations. For transportation vehicles other than autos, the indirect energy comes from similar manufacturing and infrastructure activities.

The energy associated with the processing, refining, and distribution of the energy used directly in the transportation sector (such as gasoline) is not included in the embodied energy. This energy can be calculated by fuel type and added to the indirect energy if one wishes. However, the energy associated with the processing, refining, and distribution of energy used indirectly by the transportation sector (such as the natural gas used by the steel industry to make structural components for autos) is included in the indirect energy figures.

For the entire transportation sector, the indirect energy use in 1977 was about 42 percent of the direct energy use. Thus the 26 percent of the U.S. energy use attributed directly to transportation is accompanied by another 11 percent (26 percent times 42 percent) of indirect transportation energy. Table 1 shows that the indirect energy varies in importance from a low of 34 percent of direct energy for autos and trucks to over 100 percent for rail. This should be taken into consideration when energy efficiency across modes of transportation is being compared.

For example, for almost all freight commodities rail is more efficient in moving a ton a given distance than is truck, (Shonka, 1980),if one assumes for a given commodity that the truck mode takes three times the amount of energy to move a ton from A to B. Without taking indirect energy into account, one would estimate that only 33 percent as much energy would be required for each ton being shifted from the truck to the rail mode. When indirect energy is also taken into account, the total energy used by rail is 52 percent as much energy as for truck (using the factors in Table 1).

Table 2 shows that the type of indirect energy used by the modes is quite varied. Equipment manufacture accounts for 48 percent of the indirect energy for all modes but a much lower 21 percent for buses. On the other hand, transportation services account for about 38 percent of indirect energy use in the aggregate but a much larger percent for bus (74 percent), water (60 percent) and rail (56 percent). Infrastructure indirect energy is very high for pipeline (52 percent) and very small (less than 5 percent) for air, bus, water, and rail. These percentages could change dramatically under certain conditions. For example, a major growth of rail use that required extensive new track and handling facilities could push the infrastructure component up substantially.

3. THE TECNET SYSTEM

Over a several year period International Research and Technology, Inc. (IRT) developed the Transportation Energy Conservation Network

TABLE 1

1977 TRANSPORTATION ENERGY USE

	Quads			Indirect as % of Direct*
	Direct	Indirect	Total	
Total	20.1	8.4	28.5	42
Auto	10.7	3.6	14.3	34
Truck	4.6	1.5	6.1	34
Air	2.0	1.2	3.2	62
Bus	.2	.1	.3	74
Water	.7	.6	1.3	88
Rail	.6	.7	1.3	107
Pipeline	.2	.1	.3	55
Other	1.1	.6	1.7	51

* Percentages calculated using numbers with more decimal accuracy than shown in columns 1 and 2.

Source: Doggett, et al., 1979.

(TECNET) model for OTP by building on the earlier Strategic Environmental Assessment System (SEAS) model (House, 1977) developed by EPA and DOE and further modified by Resources for the Future.

Much of the Resources for the Future work has been incorporated into the structure of TECNET. However, a considerable amount of additional detail has been added to explicitly analyze the transportation sector. These additions include a preprocessor, through which the transportation researcher can specify a host of variables to simulate alternative transportation policies; a comprehensive transportation module that estimates activity levels (i.e., passenger miles, ton miles, and vehicle miles traveled), energy use, and pollution emissions by seven modes of travel; and a set of algorithms for estimating indirect energy consumption by the transportation sector.

The TECNET system provides some unique capabilities for analyzing transportation policy. In addition to providing estimates of direct energy use (i.e., energy consumed during the operation of vehicles), the model provides estimates of energy used indirectly in support of transportation activities as mentioned briefly above. Indirect energy use is estimated for each mode (auto, truck, bus, air, water, rail, and pipeline) and is further disaggregated into three basic categories within each mode:

Transportation Equipment Manufacture

This category includes all of the energy consumed in the process of producing finished vehicles from raw materials. It includes the coal used to make steel for use in automobiles, and the electricity used by metal stamping machines and along assembly lines. Also included in

TABLE 2

ALLOCATION OF INDIRECT ENERGY USE AMONG THREE TYPES OF INDIRECT ENERGY
(Percentages)

	Equipment Manufacture	Transportation Services	Infrastructure	Total
Total	48	38	14	100
Auto	50	28	22	100
Truck	40	45	15	100
Air	49	50	1	100
Bus	21	75	4	100
Water	36	60	4	100
Rail	41	56	3	100
Pipeline	0	48	52	100
Other	100	0	0	100

EXAMPLES OF THREE TYPES OF INDIRECT ENERGY

Equipment Manufacture
- coal used to make steel used in vehicles
- electricity used in stamping machines
- electricity embodied in aluminum used in vehicles

Transportation Services
- energy used by insurance companies for lighting, heating, machine operation, paper, etc.
- warehouse energy use
- airline baggage handling equipment
- maintenance of vehicles

Infrastructure
- construction of highways, railroads, airports, dock facilities, pipelines, etc.
- maintenance of highways, railrods, airports, dock facilities, pipelines, etc.

Source: DOggett, et al., 1979

this category are estimates of energy embodied in the capital goods required to manufacture transportation equipment.

. Transportation Services

This category includes energy consumed in support activities for transportation services, such as energy used by the insurance and repair industries, and warehouse energy use. Also included in this category are estimates of energy embodied in the capital equipment used in support of transportation activities, such as airline baggage handling equipment.

. Transportation Infrastructure

This category includes the energy used in the construction and maintenance of highways, railroads, airports, dock facilities, and pipelines.

TECNET also provides detailed estimates of the economic impacts of alternative policies affecting direct and indirect energy use by the transportation sector. These estimates are provided through manipulations of TECNET's macroeconomic forecasting system, INFORUM (Interindustry Forecasting Model of the University of Maryland). INFORUM generates forecasts of the dollar output and employment by each of 185 agricultural, mining, manufacturing, and service industries. Estimates of aggregate economic statistics -- Gross National Product and its components -- are also generated in INFORUM.

In addition, TECNET provides estimates of both mobile source and stationary source pollution emissions. Mobile source emissions are estimated for each of six transportation modes. Estimates are made for hydrocarbon, carbon monoxide, nitrogen oxides, sulfur oxides, lead, and particulate emissions. Stationary point source emissions of these pollutants plus emissions of water pollutants are also estimated using the estimates of industrial activity from INFORUM.

In summary, TECNET is a comprehensive transportation research tool that makes detailed estimates of total energy use by the transportation sector plus associated pollution emissions and economic activity.

4. SYSTEM OVERVIEW

TECNET comprises six interdependent modules, each of which performs a specialized function in estimating the impacts of alternative policies for energy conservation in transportation activities. The core of the system, as shown in Figure 1.1, is the Interindustry Forecasting Model of the University of Maryland (INFORUM) (Almon, et.al. 1974). This module, developed by Professor Clopper Almon, Jr., makes detailed projections of Gross National Product, its components, and the dollar outputs of 185 agricultural, mining, manufacturing and service industries. INFORUM projections are employed by the TRANS module to estimate the direct energy requirements of the transportation sector, and by the RESGEN and ENERGY modules to estimate the generation of pollution emissions, and energy use by the non-transportation sectors of the economy. The EMBODEN (for EMBODied ENergy) module also employs projections by the INFORUM module, and combines them with embodied energy coefficients developed at the University of Illinois

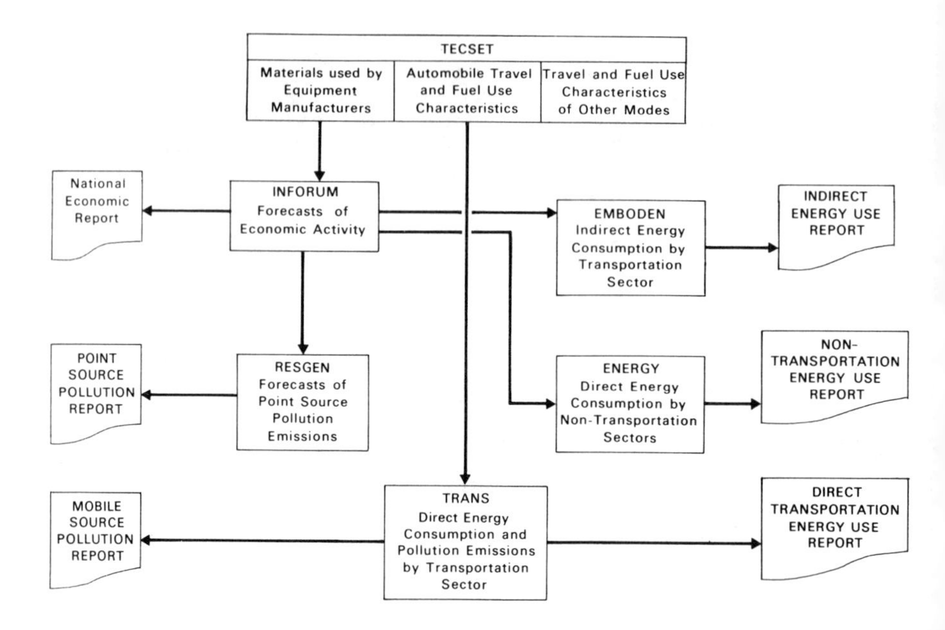

Figure 1. Simplified Diagram of TECNET.

(Herendeen and Bullard, 1974) to estimate the indirect energy requirements of the transportation sector. The TECSET preprocessor feeds user-specified scenario parameters such as material substitution trends in transportation equipment manufacturing, and travel and fuel use characteristics to INFORUM and TRANS each time the system is run.

The TECNET system provides a wide range of options from exogenous overrides to internal calculations. These capabilities allow the user of the system to specify unique scenarios to simulate the energy, economic, and environmental consequences of a given policy, or combination of policies, affecting the total energy requirements of the transportation sector out to the year 2025. Sensitivity analysis may be performed by specifying a range of values for key variables in the system. These variables include demographic projections, government expenditures, disposable income, timing and stringency of federal pollution abatement standards, materials use in manufacturing, and the rate of penetration for technological innovations, to name a few.

Direct energy demand by the transportation sector is affected most strongly by the level of economic activity. Increases in the output of certain industrial sectors will lead to an increase in the demand for freight transportation. Increases in the level of disposable income are likely to result in a higher demand for automobiles and increased automobile use. However, various behavioral, institutional, and technological changes may also have an impact on direct energy demand. A shift away from truck to rail for freight transportation is a prime example. The average truck requires nearly 4800 BTU of direct energy to move a ton of freight one mile, while a ton-mile of freight by rail requires only about 700 BTU of fuel. Similarly, a shift from automobiles to mass transit systems for urban transportation can result in a saving of 2625 BTU per passenger mile. Shifts of this kind can be specified by the TECNET user. In addition, the user can simulate direct energy conservation by altering the BTU per vehicle mile, ton mile or passenger mile coefficients to reflect improvements in vehicle performance, load practices, etc.

Indirect energy demand by the transportation sector may also be tested for sensitivity to key variables in the TECNET system. In the preliminary scenarios, it has been determined that assumptions of the level of conservation being practiced by the manufacturing sector have a major effect on embodied energy consumption. A user may also simulate the effect of changes in the materials composition of transportation equipment, as well as changes in the useful life of transportation equipment. On the energy supply side, technological improvements that reduce the amount of conversion loss associated with energy transformation may be introduced. The TECNET user can alter the efficiency of electric power generation, coal liquefaction and gasification, and the conversion of shale oil over time, frequently through assumed penetration of specific production processes.

The TECNET system initially provides the user with a "default" or "Base Case" scenario. This allows the user to specify only one or a few key parameters that will then interact with other Base Case variables to indicate the overall impacts of the changes that have been introduced. The system retains flexibility to changes in the assumptions for the Base Case, in response to changes in Department of

Energy expectations, etc. A unique scenario is constructed by changing assumptions of economic growth, future transportation activity, conservation initiatives in and out of the transportation sector, and environmental policy. The key scenario parameters in each of these areas are as follows:

. Economic and Demographic Variables

The driving force behind the INFORUM projections is the assumption of disposable income per capita in future years (i.e., the after-tax income of all persons, divided by the total population). This estimate combined with assumed levels of population, number of households, labor force, and government spending, is used by INFORUM to estimate aggregate economic activity in the future. These estimates are used in turn to estimate the outputs, employment, and investment by each of the 185 industrial categories.

Since the INFORUM model was not designed to explicitly estimate energy consumption and pollution emissions, some of the 185 sector outputs are further disaggregated in TECNET by process or by product. For example, steel production is categorized by three processes -- open hearth, basic oxygen, and electric arc, while the chemicals industry, a major source of pollution emissions, is disaggregated into over 30 chemicals. The splits by process and by product may be specified by the user, affecting the model's estimates of industrial energy use and pollution emissions.

. Transportation Activity Variables

Total demand for passenger travel, measured in passenger miles traveled (PMT), is estimated in TECNET as a function of the Gross National Product estimates generated by INFORUM. This demand is broken down into two components: travel within Standard Metropolitan Statistical Areas (i.e., SMSA or urban travel), and travel outside of and between SMSA's (i.e., non-SMSA or rural travel). Total SMSA and non-SMSA PMT's are then broken down by mode (auto, bus and rail in SMSA's; auto, bus, rail, air and water in non-SMSA's) and combined with occupancy ratios (i.e., passenger miles traveled per vehicle mile traveled) and energy intensities (i.e., BTU's per vehicle mile) to estimate total energy use by each passenger mode. Hence the user can affect passenger transportation energy use by changing any combination of the following variables:

(1) Passenger miles traveled per dollar of GNP,
(2) The split between SMSA and non-SMSA travel,
(3) Modal shares of both SMSA and non-SMSA travel,
(4) Occupancy ratios at the SMSA and non-SMSA level, and
(5) Energy intensities of passenger travel.

Because the automobile mode accounts for such a large fraction of total transportation energy use (about half in 1972), additional detail is provided in TECNET. A user may specify the shares of small, medium and large size automobiles in the new car market. A user may also specify market penetration rates for six alternatives to the internal combustion engine automobile: diesel, Brayton single-shaft, Brayton free turbine, Rankine, Stirling, and electric. In addition, TECNET contains an automobile scrappage function for more detailed analysis of the automobile stock. This function is sensitive to a user-specified assumption of the average life of automobiles, and provides for more accurate estimates of fleet average miles-per-gallon efficiency and vehicle miles traveled per automobile.

Materials use in automobile manufacturing can also be specified by the TECNET user. Changing compositions of future automobiles by size, class and engine type have a measured effect on indirect energy use. TECNET provides its users with the capability of explicitly specifying the amount of steel, aluminum, and plastics used in the manufacture of each engine type vehicle. In addition, changes in the use of all other materials, including other metals, glass, and batteries can be simulated implicitly in the economic model.

Freight transportation activity levels (i.e., ton miles traveled) are estimated as a function of the dollar outputs of INFORUM. Truck, rail, water, air and pipeline ton miles traveled are estimated at the national level and broken down into SMSA and non-SMSA as appropriate. Load factors are used to translate ton miles traveled into vehicle miles traveled, and the latter is multiplied by energy intensities to estimate total energy use. As for passenger travel, the TECNET user may simulate alternative freight transportation scenarios by changing any or all of the following variables:

1) Ton miles traveled per dollar output of all major industries purchasing transportation,
2) Splits between SMSA and non-SMSA freight transportation,
3) Modal splits at both the SMSA and non-SMSA level,
4) Load factors, and
5) Energy intensities

. Environmental Variables

In addition to changes in process and product mixes, the TECNET user can affect stationary source pollution emissions by altering assumptions regarding the stringency and timing of pollution abatement standards for industry. Mobile source emissions are affected by assumed standards and their timing in much the same way. However, these estimates are also affected by the composition of the automobile fleet which is determined using a scrappage function.

Direct energy use is associated with the vehicle miles and ton miles of travel based upon inputs by the TECNET user with respect to the fuel economy of the vehicles of the various modes over time. Thus the user can see the direct energy associated with various automobile scenarios by using internal combustion engine vehicles in one scenario, electric vehicles for part of the fleet in another scenario, a mix of turbines and stirlings in another scenario, etc.

As a result of using an input-output model, TECNET can show the inpact of material substitutions in automobiles and other transportation vehicles very easily. In fact, the user of TECNET can directly input the mix of steel, aluminum, and plastics in future automobiles via a routine made expressly for that purpose. With a little additional effort the rubber, glass, ceramics, etc. going into future autos can also be changed.

5. IMPACT ASSESSMENT

The TECNET model has been run many times during its several years of evolution to assess the direct and indirect energy impacts of various conservation measures. Since TECNET uses a comprehensive input-output model at its core, the economic and employment impacts can also be assessed but these assessments will not be reported here. Likewise, the pollution module that is part of TECNET and which generates air emissions (direct and indirect) will not be discussed in this paper. The economic,employment, and emission assessments are reported elsewhere, (Dottett, et al., 1979).

My intention in this part of the paper is to illustrate some of the broad energy impacts that we have found in having IRT run the model with our inputs. The base case was not necessarily the same in all the impact runs, so to minimize inconsistencies I will deal with percent changes in energy use between the base case and the given conservation scenario. The base case and all other scenarios do assume, however, that autos reach at least a 27.5 mpg value for new cars in 1985 and for all autos by 1995.

Table 3 shows percent changes in direct and indirect energy from a base for six alternative scenarios:

1. Conservation. This scenario combines large technology change in the auto sector (almost complete shift in new car sales to turbines, Stirlings, and electrics by the year 2000) with substantial conservation in the other modes as well.

2. Non-Auto Conservation. This scenario omits the auto technology change of the conservation case but contains all the same conservation for the other modes.

3. Gas Turbines. Gas turbines are used in all new cars by the year 2000. This is not very realistic but is done to illustrate a maximum impact case.

4. Stirling. Stirling engines are used in all new cars by the year 2000. This scenario is likewise unrealistic, but employed for illustrative purposes.

5. Electric. Electric vehicles are used in a third of new cars by the year 2000. This is also probably unrealistic, but a penetration of electric and hybrid electrics of half this magnitude is potentially possible.

6. Shift to Transit. This scenario shows the effects of mass transit increasing its ridership about 133 percent.

With regard to direct transportation energy savings, the Conservation scenario has the largest savings (26 percent) whereas the Shift to Transit scenario makes only a small dent in savings (1 percent). The greatest direct automobile savings comes from the

TABLE 3

CHANGES FROM THE BASE CASE, 2000

	Conservation	Non-Auto Conservation	Gas Turbine	Stirling	Electric	Shift to Transit
New Car sales, Year 2000						
ICE %	5	100	0	0	67	100
Turbine or Stirling	80	0	100	100	0	0
Electrics	15	0	0	0	33	0
Conservation in Other Modes	Yes	Yes	No	No	No	*
Direct Energy	-26	-18	- 8	- 9	- 2	-1
Automobile	-21	---	-22	-25	-17	-6
Truck	-36	-35	0	0	0	0
Bus	0	0	0	0	0	133
Rail	-15	-15	0	0	2	8
Air	-22	-22	0	0	0	-5
Water	-36	-27	0	0	0	0
Pipelines	-50	-50	-2	-2	-7	0
Indirect Energy	-25	-32	-2	2	3	1
Auto	-19	-21	-3	3	5	-3
Direct and Indirect	-25	-23	-6	-5	-1	0

Conservation has 20% weight reduction in average car and 165% increase in use of Aluminum.

* Shift has big shift to mass transit.

Source: Doggett, et al., 1979 and Dottett, et al., 1974.

Stirling scenario and the smallest again from the Shift to Transit scenario.

The results are quite different for indirect energy. The largest indirect transportation energy savings (-32 percent) comes in the Non-Auto Conservation case and three scenarios show an indirect energy increase! The same pattern holds for indirect auto energy use except in the Shift to Transit case where indirect auto energy is saved. The reduction of indirect auto energy use in the Shift to Transit scenario is just not enough to overcome the extra energy needs to support the growth in mass transit.

6. INDIRECT ENERGY TRENDS

The first section of Table 4 shows the annual production energy as a percent of annual direct energy for the total transportation sector and for three major modes for 1977 and 2000. Note that the percentage of production energy grows 58 percent over the period, whereas auto production energy nearly doubles and air production energy shows no increase.

The large increase in the share of auto production energy is due to the fact that direct energy goes down even in the Base Case as a result of the increasing fuel economy of autos. Simultaneously, autos are being produced using a larger amount of lighter but more energy intensive materials such as aluminum and plastics in place of steel.

The second section of Table 4 shows total indirect energy as a percent of direct energy for the entire transportation sector and the three sub-sectors that use the most energy. Note that the patterns of change over the 1977-2000 period is very similar to the first section of Table 4. The air mode has the highest percentage of indirect energy but the expected change over the period is very small.

7. CONCLUSION

Indirect energy use in the transportation sector is significant for all modes, reaching a value equal to direct energy for some of the modes. Indirect energy as a percent of direct energy is likely to grow for most transportation modes, especially for autos. Many different technology and policy scenarios can be envisioned that will reduce direct transportation energy. It is important to also consider what impact these scenarios have on indirect energy use.

TABLE 4
INDIRECT ENERGY TRENDS TO THE YEAR 2000, BASE CASE

	1977	2000	Percent Change
Production of Vehicle Energy as a Percent of Operating Energy in Given Year			
- Transportation Sector	20.0	31.6	58
Auto	16.6	32.4	95
Truck	13.8	17.5	27
Air	30.0	30.1	0
Total Indirect Energy as a Percent of Direct Energy			
- Transportation Sector	41.9	60.3	44
Auto	33.5	58.3	74
Truck	33.6	38.9	16
Air	61.7	64.2	4

Source: Doggett, et al., 1979

BIBLIOGRAPHY

Almon, Clopper, et al., 1985: Interindustry Forecasts of the American Economy, Lexington Books, (1974).

Doggett, Ralph, et al., Further Development and Use of the Transportation Energy Conservation Network (TECNET), International Research and Technology Corp. for the Department of Energy, HCP/M2101-2, March, (1979).

Dottett, Ralph, et al., Ten Scenarios of Transportation Energy Conservation using TECNET, International Research and Technology Corp. for the Department of Energy, HCP/M2101-1, March, (1979).

Herendeen, Robert and Clark Bullard, Energy Cost of Goods and Services, CAC Document No. 140, Center for Advanced Computation, University of Illinois, Urbana-Champaign, 6 Novembver, (1974).

House, Peter, Trading off Environment, Economies, and Energy: A Case Study of EPA's Strategic Environmental Assessment System, Lexington Books, Lexington, MA, (1977).

Kulp, Gretchen, editor, Transportation Energy Conservation Data Book, Edition 4, Oak Ridge National Laboratory, (Forthcoming).

Shonka, Debbie, Transportation Energy Conservation Data Book, Edition 3, Oak Ridge National Laboratory, ORNL-5493, February, (1979).

Users Guide for the Transportation Energy Conservation Network TECNET), International Research and Technology Corp. for the Department of Energy, June, (Draft report), (1980).

13 Integrated Environmental Models: A Look at Some Diverse User Needs

DAVID A. BENNETT AND JOHN W. REUSS

1. INTRODUCTION

The decade of the 1980's poses great challenges for environmental policymakers. These challenges are serious, real and generally well recognized. Among them is the emergence of a national energy policy emphasizing dramatic increases in the use of coal. Federal policies enacted to stimulate economic growth by "reindustrializing America" have important environmental implications. Likewise, the continuing debate over the regulatory process as well as the substance of regulatory policies may mean agencies like Environmental Protection Agency (EPA) will have to devote a greater share of their limited resources to defending their regulatory base at the cost of developing data and tools required to meet tomorrow's environmental threats and insults.

Historically, regulatory agencies have always been subject to criticism, scrutiny, and stress. What seems to mark the 1980's is increased acrimony which results in regulatory agencies spending greater portions of limited resources defending the status-quo. This posture means fewer resources than ever available for longer term considerations. This new mood puts new emphasis on our political system's already heavy preoccupation with the short-term and makes the task of the R&D planners and analysts responsible for assessing the long-term future more difficult and tenuous.

It is against this background that we wish to describe one effort to provide information on long-term environmental problems and opportunities to R&D policymakers, our use of and experience with a large computer model, and our judgments about the kinds of analytical tools we need to accomplish our objectives.

2. THE USER - OFFICE OF EXPLORATORY RESEARCH

EPA's R&D program is intended to support the Agency's regulatory program and to look "over the horizon" to provide information about potentially significant future environmental problems. The Office of Exploratory Research is a recent addition to R&D at EPA that provides "over the horizon" capabilities.

When in 1970, fifteen agencies from five departments were combined to create EPA, the highest, most immediate priority was to deal with highly visible air and water quality problems. Although it was recognized that the Agency had to have a R&D capability for its science and technology based regulations to be defensible, dealing effectively with these initial high-priority problems was not considered to be primarily a R&D problem. In part, this was because there was a great deal of applicable on-the-shelf knowledge that recent environmental legislation and the new agency would make it possible to bring to bear.

Things have changed significantly since then. For one thing, the Congress has greatly expanded EPA's legislative mandate, including adding problems, such as the control of toxic and hazardous substances that are more difficult to deal with. If it ever was, on-the-shelf knowledge is no longer adequate. And many of the environmental problems the Agency must now attempt to deal with are basically R&D problems. This has led to major changes in the Agency's R&D program and planning process.

A part of the impetus for change has come outside EPA, particularly from the Congress and the National Academy of Sciences. Both organizations recognized the dual needs of supporting the Agency's regulatory decision making and anticipating future environmental problems.

The Environmental Protection Agency responded in 1977 by changing its R&D planning process by establishing thirteen intra-agency Research Committees and an Anticipatory Research Program and a Strategic Analysis Group. The last two have since been combined to form the Office of Exploratory Research. Figure 1 illustrtes EPA's new R&D planning process and shows where the Research Committees and Office of Exploratory Research fit into the process.

As indicated in Figure 1, the Agency's basic R&D plan is the annual Research Outlook report. The Congressional mandate calls for this comprehensive five-year environmental R&D plan to be delivered annually to the Congress no later than two weeks after the President submits the administrative budget.

Inputs into the plan come from several sources: research strategy documents prepared by each of the thirteen intra-agency Research Committees; advice from the Agency's Science Advisory Board; recommendations from studies performed for the Agency by the National Academy of Sciences and others, and the results of studies conducted by the Office of Exploratory Research, including findings reported in an annual Environmental Outlook report. And, as Figure 1 indicates, the plan is shaped by the budget process and the results of completed and current research.

The Office of Exploratory Research is responsible for providing both anticipatory and exploratory inputs into the R&D planning process by identifying, defining and assessing future environmental trends and problems and conducting exploratory research to meet basic knowledge needs concerning potential future problems. Within Exploratory Research, the Office of Strategic Assessment and Special Studies is responsible for strategic analysis, including forecasting and assessing future environmental trends and problems. The overall objectives of Strategic Assessment are:

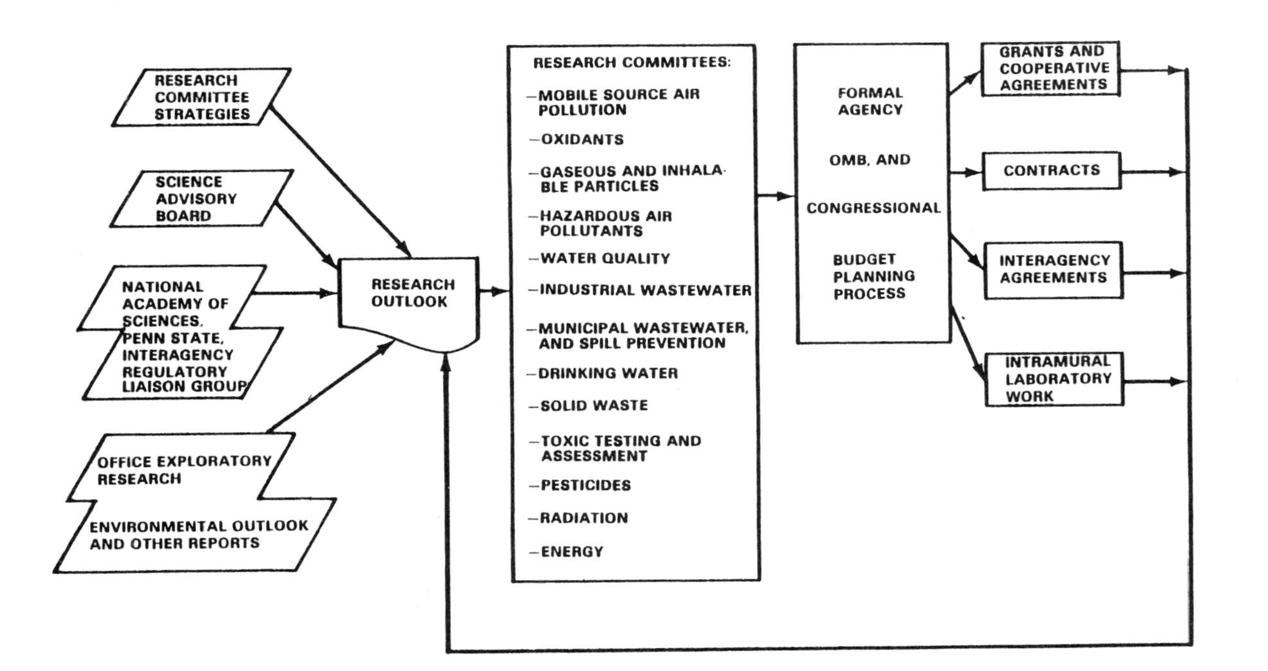

Figure 1. EPA's R & D Planning Process.

1. Developing a system for identifying and assessing the significance of future environmental problems and opportunities, including:
   (a) environmental indicators and indices for detecting and "measuring" changes in environmental quality; and
   (b) participation mechanisms which tap both expert knowledge and public perceptions;
2. Improving the credibility, quality and utility of strategic analysis products as an input into R&D and Agency planning; and
3. Communicating effectively its findings on future problems with other parts of the Agency.

The program for achieving these objectives consists of five major, closely related activities: Environmental Outlook, Formal Analytical Methods, Outreach Methods and Activities, Mini-Assessments and Special Studies. The relationship of these five activities to each other and to participants in EPA's R&D planning process is depicted schematically in Figure 2. Briefly described, these activities and their functional purposes are:

- Environmental Outlook: As Figure 2 shows, Environmental Outlook is the core organizing strategic analysis activity. Reports produced as part of this activity, including an annual _Environmental Outlook_ report, are an important means of communicating information about the environmental future to participants in the R&D planning process. These participants are the members of the thirteen Research Committees responsible for preparing R&D strategy documents, the persons responsible for preparing the annual _Research Outlook_ plan, and the persons who participate in developing the Agency's budget. These are not mutually exclusive categories since some persons are involved in all three aspects of R&D planning.

  Formal Analytical Methods and Outreach Methods and Activities: Together these two activities are responsible for developing the capability to provide a comprehensive, synthetic _Environmental Outlook_. The Formal Analytical Methods activity is intended to develop and maintain the requisite formal strategic analysis methods and techniques. The SEAS model is one analytical tool maintained. We intend to develop others. Outreach Methods and Activities is intended to develop and maintain methods and techniques for obtaining citizens' perceptions and experts' opinions about future environmental trends and problems; it is also responsible for introducing them into the _Environmental Outlook_ reports and for securing their participation in its planning and preparations.

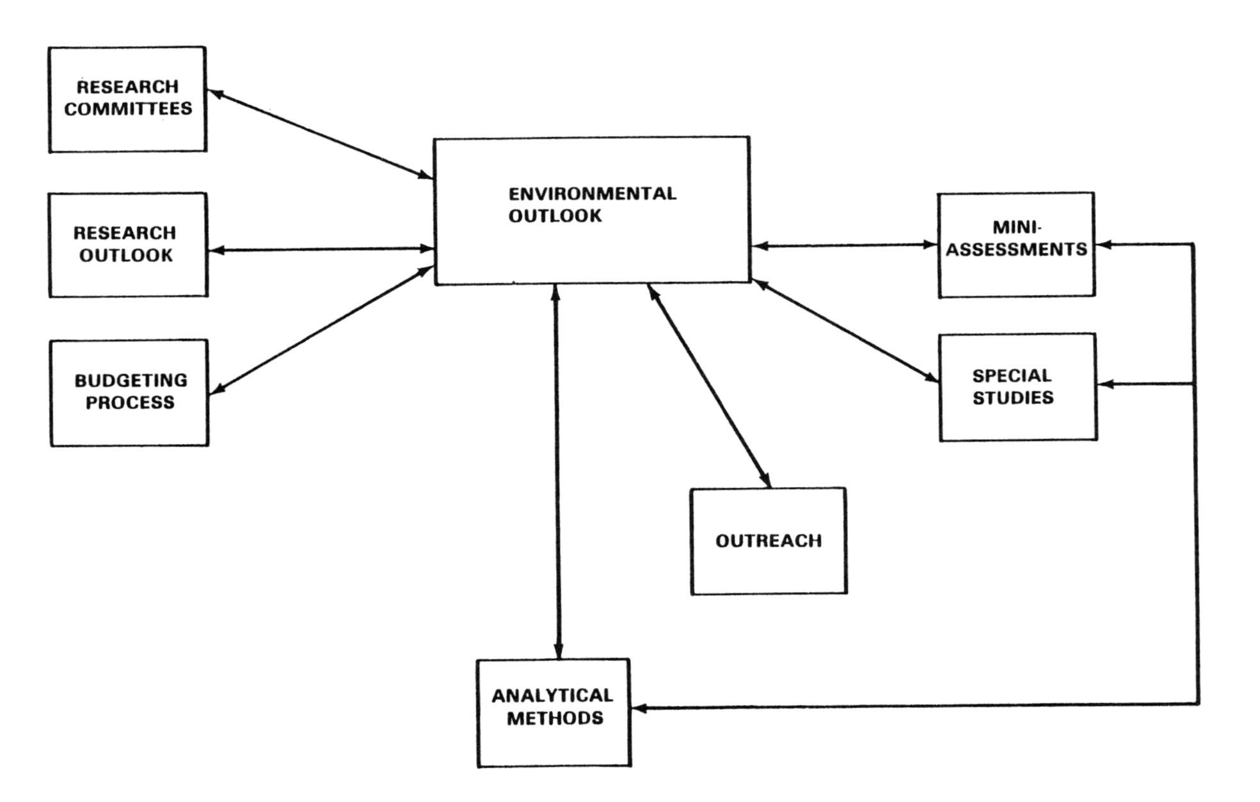

Figure 2. The Office of Strategic Assessment and R&D Planning.

- Mini-Assessments and Special Studies: Some future environmental concerns and potential problems identified by the Environmental Outlook activity will require further definition and analysis to determine whether they are likely to be problems, and, if they are, whether they are likely to be serious enough to justify further research. The Mini-Assessment and Special Studies activities are intended to provide the information needed to answer these questions. The least well-defined concerns and potential problems are candidates for mini-assessments, which are small-scale problem identification and definition studies. Problems about which more is known but for which still more knowledge is needed to be able to identify research needs are candidates for special studies. Special studies also include large, integrated assessments, for example, of a region, an industry, or a particular resource. These applied policy studies are intended to integrate results of EPA's overall research programs, regional offices and others interested in large-scale and cross-cutting problems. They are also intended to help ORD identify gaps in its existing R&D program.

Two major projects of our office that can usefully utilize integrated environmental models are the Environmental Outlook reports and large-scale integrated regional technology assessment studies.

3. ENVIRONMENTAL OUTLOOK 1980

Our first attempt to produce a reasonably comprehensive overview of the future environment has just been published. The Environmental Outlook 1980 report attempts to provide historical trends information on public opinion about environmental protection and environmental policy, and projects future trends for a number of pollutants and analyzes their environmental implications. (See Table 1). This report, based largely on SEAS projections, was produced by an interdisciplinary team comprised of persons on the staff of the Strategic Analysis Group and its support contractors. An executive summary will soon be published.

Obviously, this report and the kind of information it contains about the environmental future have policy implications beyond R&D planning. For example, the analysis of future trends includes making assumptions about environmental regulations. In our 1980 report, our projections based upon SEAS scenarios indicate that existing regulations for controlling air pollution, if fully complied with, would result in a decline in the national emissions of several so-called conventional air pollutants between 1975 and 2000. However, we also conclude that emissions of sulfur oxides and nitrogen oxides would increase, nitrogen oxides significantly. The major factors producing these trends are the projected increased use of coal by electric utilities and industrial boilers and the controls imposed on point and non-point and mobile sources of air pollutants. Our projections also show that non-point sources of water pollution, such as agricultural and urban runoff, would greatly exceed the discharges from point sources, such as industrial plants, and would affect 90 percent of the drainage basins in the U.S. If not

TABLE 1: CONTENTS OF ENVIRONMENTAL OUTLOOK 1980

- Societal Trends
- Air Pollutants
- Global Atmospheric Pollution
- Water Pollutants
- Drinking Water
- Water Resources
- Ocean Pollution
- Solid and Hazardous Wastes
- Radiation
- Noise
- Energy and the Environment

---

adequately controlled, which they now are not, this non-point pollution would prevent national water quality goals from being achieved.

Neither of these findings will surprise the responsible EPA program office. But the analysis will be useful to them and to the persons in the Agency responsible for providing guidance when priorities are set and resources are allocated. That is why, although our immediate responsibility is to provide environmental futures information for R&D planning, we devote so much of our strategic analysis effort to working closely with the program and planning and management offices. We intend our applied policy and analysis products, such as the Environmental Outlook, to be a major policy planning resource. For these products to provide a basis for well-informed R&D planning, it is essential that they serve this broader purpose as well.

Although the Environmental Outlook is an annual report, we will not attempt to present a comprehensive overview each year. Rather, we will do this about once every three or so years. In the in-between years, the report will consist of an updated summary of the most recent data on future pollutant loadings, and several chapters which focus on problems which were identified as special attention. For example, Environmental Outlook 1981 will include chapters which focus on a number of social trends (with an emphasis on detailed demographic profiles and life styles), wet and dry chemical deposition, changing chemical feedstocks; and agricultural practices.

By 1983, the 1981-1982 Environmental Outlook reports, and the Formal Analytical Methods, Outreach Methods and Activities, Mini-Assessments and Special Studies accomplishments will make it possible to produce a much more comprehensive, synthetic overview of the environmental future than we were able to produce for 1980. To provide this overview, we must develop or adapt environmental modeling capabilities.

4. INTEGRATED ENVIRONMENTAL ASSESSMENTS

Another activity of our office is the design and management of integrated environmental assessments to evaluate future impacts of regional changes such as large-scale development of energy resources or significant demographic or economic shifts. One requirement for a successful assessment is a regional forecasting model that is transparent and provides base year and short-term projections closely resembling "ground truth" recognized by regional participants in the assessment activity.

The regional model would provide a forecast of feasible futures using either bounding scenarios or some form of contingency analysis as a basis for identifying potentially significant environmental problems during the next 25-30 years. Special attention must be given to:

- basic driving forces such as population, economic growth, energy and raw materials development;
- new technology;
- wider scale or more intensive use of established technologies;
- conversion of land uses, particularly loss of wetlands and the conversion of agricultural land to other uses;
- changes in agricultural technologies and practices; and
- limits on available natural resources.

To provide all this and the "ground truth" too is the modeler's challenge.

5. THE NEEDS - REGULATORY ANALYSIS AND ENVIRONMENTAL FUTURES RESEARCH

Two potential modeling applications -- Environmental Outlook and Integrated Environmental Assessment -- were discussed above. Here we identify details and needs that are often asked of the Agency's R&D to support its regulatory mandate. As environmental laws have become more complex and comprehensive, R&D has received increasingly complicated requests, and the temptation is to expand available models to answer these requests. Each request can be viewed as identifying a need to the model. It is probable that no one model, or interlocking group of models, can meet all of these requests. Nevertheless, we need to ask, how appropriate are integrated environmental models in addressing these needs?

Four needs are identified below:

1. Models must go beyond the estimation of residuals.

   Outputs from the SEAS model are in tons of residual emitted (e.g., SO_2 to air) per year for a specified region. Although emissions information is valuable and forms the basis for much environmental regulation, ultimately environmental laws must be justified as protecting public health or welfare. Although the Clean Air and Clean Water Acts emphasize emissions and "end of the pipe" technologies to control emissions, they also address ambient air and water quality standards. It is ambient air and water quality that affects human health and welfare.

For individual pollutants it may not be necessary, or even desirable that integrated models include modules that translate emissions to ambient levels. However, a growing concern is synergistic interactions of pollutants in the ambient environment. Three examples in air are: (a) the respiratory effects of sulfur dioxide and particulates; (b) the interactions of particulates and sulfur oxides and nitrogen oxides to yeild acidic precipitation; and (c) the roles of nitrogen oxides and hydrocarbons in the formation of photochemical smog and the impairment of visibility. Therefore, at least some modeling of interactions is necessary.

2. Models must address increasingly complex environmental laws.

Although Federal environmental laws have existed since the Rivers and Harbors Act (or Refuse Act) of 1899, the bulk of EPA's regulatory authority resides in laws promulgated in the 1970's. In 1970 there existed laws addressing a handful of pollutants discharged to the air, water, or land media. Transformation of pollutants and transfers between media were generally not addressed. Today the complement of environmental laws considers hundreds of chemical substances, alone and in mixtures, from their production to their ultimate disposal. Objectives and approaches of major environmental laws are summarized in Table 2.

Many of the differences in approaches to the control of environmental pollutants are related to the type of substances regulated -- residuals from industrial processes, energy production, and waste treatment systems -- under the Clean Air Act, the Clean Water Act, the Resource Conservation and Recovery Act, and the products themselves under the Toxic Substances Control Act (TSCA) and the Federal Insecticide, Fungicide, and Rodenticide Act (FIFRA).

The Clean Air Act and the Clean Water Act require technology-based levels of emission control and include ambient air and water quality standards. The Resource Conservation and Recovery Act sets up a system for the management of pollutants rather than prescribing the level of pollutants to be attained in any medium. However, the standards for management must be sufficient to protect human health and the environment, a test that is similar to those used in setting ambient and effluent standards under the other Acts. TSCA and FIFRA control the testing, production, and use of hazardous and toxic products through regulations.

Several of the Acts consider specific pollutants. Table 2 cites the criteria and hazardous pollutants of the Clean Air Act and the conventional pollutants of the Clean Water Act. In addition, a list of 129 toxic pollutants was included in Section 307 of the Clean Water Act as a result of a 1976 settlement in a U.S. District Court.

This suggests that modeling releases of all pollutants to all media would be most difficult. Toxic substances are

TABLE 2

OBJECTIVES AND APPROACHES FOR CARRYING OUT OBJECTIVES OF MAJOR ENVIRONMENTAL LEGISLATION

Environmental Component Affected	Most Recent Legislation	Objectives of Legislation	Types of Pollutants Considered	Approach and Requirements for Carrying Out Objectives[a]
Air Quality	Clean Air Act Amendments of 1977 PL 95-95	° Protect and enhance quality of air resources to promote public health and welfare ° Establish national research and development program for prevention and control of air pollution ° Provide assistance to states for air pollution control programs	° Criteria pollutants sulfur dioxide particulates carbon monoxide photochemical oxidants hydrocarbons nitrogen oxides lead ° Hazardous air pollutants asbestos beryllium mercury vinyl chloride	° National Ambient Air Quality Standards (NAAQS) [109] ° State Implementation Plans (SIPs) [110] ° National Emission Standards for Hazardous Air Pollutants (NESHAPs) [112] ° New Source Performance Standards (NSPS) [111] ° Emission standards for new motor vehicles or motor vehicle engines [202]
Water Quality	Clean Water Act of 1977 PL 95-217	° Restore and maintain the chemical, physical, and biological integrity of the nation's waters by: ° Eliminating pollutant discharge into navigable waters by 1985 ° Achieving water quality suitable for protection and propagation of aquatic life and water recreation ° Prohibiting the discharge of toxic pollutants in toxic amounts	Specific pollutants discharged into water [502(6)] including: ° Conventional pollutants (BOD, suspended solids, fecal coliform, and pH) [304(a)(4)] ° Nonconventional pollutants (neither of the above) [301(2)(F)]	° New Source Performance Standards (NSPS) [306(b)(1)] ° National Pollutant Discharge Elimination System [NPDES] [301(b)(2)] ° Combination of technology standards (e.g., "best practicable control technology," [301(a)(1)(A)], effluent limitations (308), and toxic effluent standards [307(a)]

(Continued)

Table 2 (continued)

Environmental Component Affected	Most Recent Legislation	Objectives of Legislation	Types of Pollutants Considered	Approach and Requirements for Carrying Out Objectives
Solid Wastes	Resource Conservation and Recovery Act of 1976 PL 94-580	° Provide technical and financial assistance for the development of management plans and facilities for the recovery of energy and valuable materials from solid waste ° Provide for the save disposal of of discarded materials ° Regulate the management of hazardous wastes	° Solid waste-discarded material from industrial, commercial, mining, and agricultural operations and from community activities [1004 (27)] ° Hazardous waste-solid waste or combination of solid wastes that may cause illness or pose a hazard to human health or the environment [1004(5)]	° State plans for disposal of nonhazardous wastes [4006] ° Manifest system through which every load of hazardous waste material can be tracked from generation to ultimate disposal [3002(5)] ° Permit program to cover every hazardous waste disposal site [3005]
Toxic and Hazardous Substances	Toxic Substances Control Act of 1976 PL 94-469	Regulate commerce and protect human health and the environment by requiring testing and use restrictions on certain chemical Substances	Chemical substances or mixtures that present or will present an unreasonable risk of injury to health or the environment [6(a)]	° Testing of chemical substances and mixtures [4] ° Notification to EPA of manufacture of new chemical or new use for a chemical [5] ° Regulation of hazardous chemical substances [6]
Pesticides	Federal Pesticide Act of 1978 PL 95-396	To prevent unreasonable hazards to humans or the environment. (implicit)	Pesticides	° Registration of pesticides and producers ° Classification for general use, restricted use, or denial of registration (unregistered pesticides may not be shipped, sold, or delivered) ° Certification of users of restricted use pesticides

[a] Numbers in brackets refer to Sections of applicable Acts.

[b] NRDC v. Train, No. 45-172, 8 ERC 2120, D.D.C., June 8, 1976

particularly troublesome, for they are often produced in very small quantities and may be formed as byproducts of other processes. Yet each substance, whether produced by the ton or by the ounce, is important under the law. Indeed, small quantities of very toxic substances may pose the greatest threat to human health and welfare. These subtleties and complexities require analysts and modelers to define their problems carefully.

3. Models must approximate "ground truth" upon disaggregation of results.

   One of the major problems of the SEAS model is that as one disaggregates industrial and energy activity from the national level to smaller and smaller regions, facilities may be sited by SEAS algorithims in locations that local experts know to be impossible. This problem is especially severe at the county level, largely, because of inconsistencies between the OBERS regional forecasts of future demographic trends and the other regionalization modules.

   The inconsistencies in the SEAS model can be remedied, but the question remains -- how far can or should a national marcoeconomic model be disaggregated? Extreme disaggregation costs are high. Development costs may be dwarfed by the potential cost of the loss of the model's credibility. At some point regional models must be more appropriate. Tasks must be assigned to the model appropriate to its design.

4. Models must be capable of accepting "non-business-as-usual" scenarios and of addressing "what if" questions.

   A modeling tool, to be useful for strategic analysis, must be accessible, flexible, and efficient. The effects of unusual energy, economic, demographic, or environmental scenarios must be readily and quickly determined.

   Our experience with the SEAS model has been that more than two month's time elapses between posing even one question and the completion of preliminary analysis. The new DOE/EPA SEAS reference system seems to run well and provide national and regional projections of conventional pollutants, but it does not meet the strategic analysis criteria. Perhaps level of detail must be given up to allow responsiveness.

This last need is of particular concern for strategic assessment, for we often work in anticipation of laws. Thus, the analysis may be less complex, but the problems are often unusual. To illustrate this point, consider the following nine mini-assessment topics identified for study in 1980:

- Social, institutional and behavioral factors in siting "undesirable" facilities such as hazardous waste facilities;
- Agricultural uses of applied genetics and biotechnologies;

- Industrial use of applied genetics and biotechnologies;
- Increased use, recycling and combustion of composite materials;
- Increased production and use of alcohol-based fuels;
- Increased use of wood and other biomass sources of energy;
- Alternate feedstocks for basic industrial chemicals;
- The current overcapacity to produce bromines and the possible alternative uses of organobromides; and
- Possible decentralization of the work-place due to developing communications technologies.

Each topic represents a possible future. The mini-assessment studies consider the nature and probability of that future. For any probable future, we need to ask "what if" questions that address a number of environmental implications. In addition, we need to be able to turn to models that are well-documented, transparent, and efficient.

6. CONCLUSION

The current economic/regulatory climate, as compared to that of a decade ago, is one that provides more complicated environmental problems deserving analysis but fewer resources to support analysis. The former may encourage the development of more complex models or the expansion of current models. To the extent that this results in a model whose linkages are opaque and is bulky and slow, it will gain few users in the environmental community. The constraint of limited resources and the needs of strategic environmental assessment clearly favor the development of streamlined models, perhaps less comprehensive in their analysis, capable of efficient analysis of contingencies.

14 Regional Demand Implications for Gasoline Supply Shortages

DAVID L. GREENE

1. INTRODUCTION

In the past decade the United States has experienced two periods of petroleum and gasoline supply shortages. Both the 1973-74 and 1979 petroleum import interruptions initiated brief periods of rapid petroleum price escalation. In the seven years since the 1973-74 Arab OPEC oil embargo no reduction in our vulnerability to possible future supply interruptions has been achieved. In 1973 the U.S. consumed petroleum at the rate of 17 million barrels per day. Of this 6.2, million barrels, or 36.2 percent, was imported and 48 percent of imports were obtained from OPEC nations. In 1980 U.S. petroleum consumption averaged 17 million barrels per day, down from 18.5 in 1979. Petroleum imports in 1980 averaged 6.8 million barrels per day or 40 percent of total domestic consumption (down from 8.5 mbd the year before). Of this, 62 percent was obtained from OPEC sources. Finally, in 1973 30 percent of our OPEC imports came from Arab members of OPEC. In 1980 the corresponding figure was 60 percent (U.S. Department of Energy, 1981).

Put in perspective, the 1979-1980 reduction in petroleum imports provides little comfort, especially since it was largely brought about by a doubling in the prices of crude oil from 1978-1980. Moreover, that gigantic price increase was consequent to a minor worldwide petroleum shortage of about 4 percent caused by the loss of supply during the Iranian revolution (Alm, 1981). In brief, it is clear that we have not reduced our vulnerability to petroleum supply interruptions since 1973 and in fact our strategic position may well have worsened.

Motor gasoline constitutes the single largest component of petroleum fuel use in the United States. Consumption in 1980 averaged just under 6.6 mbd, slightly lower than in 1973. In any petroleum supply emergency, gasoline supplies are certain to be disrupted with serious consequences for the economy, substantial inconvenience to the public, and serious potential threat to national security. Past shortages of gasoline in 1973-1974 and summer 1979 have been relatively minor in comparison to the potential disruptions we now face. Gasoline shortages at the national level in 1979 have been estimated at no worse than 4 percent (Greene and Chen, 1981). The Emergency Energy Conservation Act of 1979 (U.S. Senate, 1979) provides for gasoline rationing only in the event of a petroleum shortage of 20%.

Several analyses have been done at the national level concerning severe petroleum shortages, and the effects of various strategies for coping with them. However, virtually no analyses of the regional dimensions of the problem have been made. The purpose of this paper is to address those issues. There are two very important geographical aspects of petroleum supply emergencies about which we are notably ignorant. The first, is the ability of the crude petroleum and petroleum product distribution system to efficiently and equitably distribute available supplies in a severe shortage. Despite the fact that our experience has shown that regional variations in shortage conditions can be extreme, practically nothing has been done to understand and solve this problem (Osleeb, 1979, is an exception). The second issue is regional variations in the nature of gasoline demand. Variations in regional price elasticities conversely imply variations in willingness to pay for additional supply during a shortage. If compounded by varying regional supply shortages the possible variation in regional impacts is even greater.

This paper examines the question of regional variability in gasoline demand responsiveness and evaluates its implications for several pricing and rationing shortage management schemes. Unless future supply shortages differ from those in the past there will be varying degrees of shortage across the U.S. creating a regional equity problem. Because of differences in gasoline demand responsiveness across regions, shortage management policies may either ameliorate or exacerbate the equity problem. Policies also differ with respect to economic and bureaucratic efficiency. The remainder of the paper is divided into two sections. In the first the variability of gasoline price elasticity across states and the implications for regional responses to supply shortages are examined. In the second a typology of price control and floating price strategies are analyzed with respect to their economic efficiency and regional equity. It is concluded that the key to solving the regional equity problem lies in insuring that available supplies are correctly redistributed.

2. SHORTAGE IMPACTS AND THE PRICE ELASTICITY OF DEMAND

In the past, petroleum supply shortages have had at least three kinds of harmful effects. Because of price controls on petroleum fuels and the absence of an efficient allocation scheme, fuels were simply unavailable for certain uses or fuels such as gasoline were rationed by queuing rather than price. This, of course, results in a deadweight economic loss in wasted time and ineffecient allocation of resources. Second, the shortages have served as a mechanism by which the oil cartel can dramatically raise oil prices. Whether the shortages served as a signal to raise prices or were deliberately used as a lever to raise prices is moot. The fact remains that past shortages revealed to producers and consumers alike the inelasticity of short-run demand for petroleum, thus creating the opportunity for large price increases. Finally, those sudden, gigantic price increases pushed the national economy out of equilibrium, making much of our capital stock of energy using durable goods obsolete and causing inflation.

At present there are no price controls on petroleum and its products, but neither is there any plan in place for dealing with a petroleum supply interruption. In the event of a shortage, fuels would be rationed by market prices, unless the government took some sudden action to obstruct it. In this case there would be no deadweight economic loss since market prices would almost certainly allocate the available resources in an efficient manor. There would, however, be a substantial transfer of wealth from the buyers to the owners of the suddenly scarce resource.

Consider the case of a 20% shortage of gasoline. If the price elasticity of demand is -0.2 and constant for all levels of demand, then prices should increase threefold.* In 1980 consumers purchased about 6.6 million barrels of gasoline per day at an average price of about $1.25 gallon in current (1980) dollars. At market prices during a 20% shortage they would pay about $3.80 per gallon for only 5.3 mbd. The total daily income transfer would be

$$(5.3 \times 10^6)\ (42)\ [3.80 - 1.25] = \$568 \times 10^6$$

or about a half billion dollars per day. On an annual basis this amounts to 207 billion dollars, about one third the size of federal government expenditures in 1980 (approximately 601×10^9, 1980 dollars, U.S. Department of Commerce, 1981), and almost as large as all federal government transfer payments in that year (250×10^9, 1980 dollars, U.S. Department of Commerce, 1981). Clearly an income transfer of this magnitude should not be permitted without due consideration.

While the dimensions of the problem at a national scale are reasonably well understood (e.g., Difiglio, 1980), very little analysis has been done concerning regional issues. The key to understanding the regional issues is knowledge of regional gasoline demand functions. Until recently, attempts to estimate regional gasoline demand equations have achieved dubious results (e.g., see Greene, 1980a and 1980b; Johnson, 1980; Kraft and Rodekohr, 1978 and 1980; Mehta, et al., 1978). State and regional price elasticity estimates have tended to vary greatly, many having incorrect signs or lacking statistical significance. Reasons for the failure to obtain reasonable regional gasoline demand models have been traced by Greene (1980a) to basic data inadequacies. Inaccuracies in state level data, brevity of the historical time series used and a lack of variation in prices have all

---

* If demand, q, is a function of price, p,

$$q = ap^{-0.2}$$

then in a 20% shortage

$$\frac{q_2}{q_1} = 0.8\left(\frac{p_2}{p_1}\right)^{-0.2}$$

and

$$(0.8)^{-5.0} = \frac{p_2}{p_1} = 3.06 \quad .$$

There is statistical evidence to support the assumption of constant elasticity. Using Box-Cox functional form analysis techniques, Danielson and Agarwal (1976) and Greene (1980) both concluded that the optimal functional form for a U.S. gasoline demand equation was not statistically significantly different from the constant elasticity form.

contributed to the inherent difficulties of constructing regional models.

Recently a revised series of monthly state gasoline consumption data has become available covering the period January 1975 to July 1980. This new series is considerably improved over previous series and for the first time permits the estimation of individual state time series models of gasoline demand for a period which includes a very large increase in gasoline price. The results of that analysis are described below.

3. STATE PRICE ELASTICITIES OF GASOLINE DEMAND

Gasoline demand models were estimated for each state and the District of Columbia using the Autoregressive-Integrated-Moving-Average (ARIMA) technique developed by Box and Jenkins (1976). Gasoline sales data were taken from the Federal Highway Administration's series MF-33G monthly motor gasoline sales by states as revised in 1981 (U.S. Department of Transportation). Input variables were national gasoline prices (Platts, 1979; U.S. Department of Energy, 1980) and monthly personal income by state, supplied courtesy of Business Week magazine. Monetary variables were deflated to constant dollars using the consumer price index for all urban workers (U.S. Department of Commerce, 1976-1980). State consumption and income were converted to per capita values by a linear-in-logarithms interpolation between 1975 and 1980 state census data (U.S. Department of Commerce, 1977 and 1981). In addition to price and income, intervention variables were introduced to account for the effects of the gasoline shortages during May, June, and July, 1979. The general functional form used for all state models is given by

$$q_t = w_1 p_t + w_2 y_t + \sum_{i=1}^{3} \mu_i z_{it} + N(u_t) \qquad (1)$$

where all variables except the z_{it} are expressed in logarithms. q_t is per capita gasoline sales, p_t the constant dollar price of regular grade gasoline, y_t is per capita personal income, z_{it} are variables which take on the value of one in a particular shortage month and are zero otherwise. $N(u_t)$ is the ARIMA noise node which includes differencing, moving-average, and autoregressive components, at least in principle, and is a function of a white noise error term u_t. Typically state noise models included a twelfth difference in the denominator and one or two moving average terms in the numerator. For example,

$$\frac{(1 - \theta_1 B - \theta_2 B^2)u_t}{(1 - B^{12}} \qquad (2)$$

is a second order moving-average noise model with one seasonal difference of order 12 (B is a backward operator defined by $B^n u_t = u_{t-n}$). The effect of putting $(1 - B^{12})$ in the denominator is to take the twelfth difference of all variables in equation (1).

The results of the state model estimations are described in detail in Greene and Chen (1981). Statistically acceptable models for all states were obtained. Table 1 summarizes the estimated state price elasticities. All state elasticity estimates are negative, as one would expect. Only three, Alaska, Kentucky, and Pennsylvania, are not significantly different from zero at the 0.05 level of significance. Such results are remarkable in comparison to previous efforts to estimate state elasticities (e.g., see Mehta, et al., 1978). Statistically significant price elasticities range from -0.138 (Ohio) to -0.377 (Iowa). The median price elasticity for states is -0.238 and the unweighted, arithmetic mean is -0.242, indicatring a fairly symmetrical distribution of elasticity estimates.

When mapped, state price elasticity estimates exhibit some interesting regional patterns (Figure 1). A belt of northeastern states from Massachusetts to Virginia and westward to Indiana and Kentucky are relatively unresponsive to price changes. All states in this region have price elasticities below average (in absolute value). Southern tier states from Florida to Texas behave much like the national average. Price responsiveness is greatest in northern Plains and Rocky Mountain states, while western states exhibit moderate to high price elasticities. Even more than the statistical significance of individual state price elasticity estimates, the existence of these readily discernible regional patterns strongly supports the existence of significant variations in the price responsiveness of gasoline demand across the United States. It is extremely unlikely that such distinct regional patterns would emerge if state-to-state variations were due simply to random state errors or were artifacts of the model structure.

No comprehensive explanation for the observed regional variability of price elasticities exists at this time. Yet there does appear to be a relationship between per capita gasoline sales and the elasticity of demand. This relationship is illustrated in Figure 2. The simple correlation between elasticity and consumption seen in Figure 2 is -0.62. A simple linear least squares regression of elasticity on level of sales per capita yielded the following relationship.

$$e_i = 0.0298 - 0.00048\ q_i \tag{3}$$

where e_i is the price elasticity for state i. At the mean elasticity of -0.24, per capita sales on the regression line are 555 gallons per year per capita. The elasticity of price elasticity with respect to consumption at the mean is,

$$\frac{de}{dq} \cdot \frac{q}{e} = -00048. \frac{555}{-0.24} = 1.12 \tag{4}$$

The implication is simply that the more gasoline consumed per capita, the greater the willingness to forego consumption when prices increase. Intuitively, this is an appealing result. The first few trips given up in short run response to a price increse will obviously be those least valued. As prices increase further, trips which are more necessary must be relinquished. Presumably the willingness to pay for these trips will be greater. If essentially the same principle applies across states as average per capita sales vary, then we would expect price elasticity to vary with level of consumption.

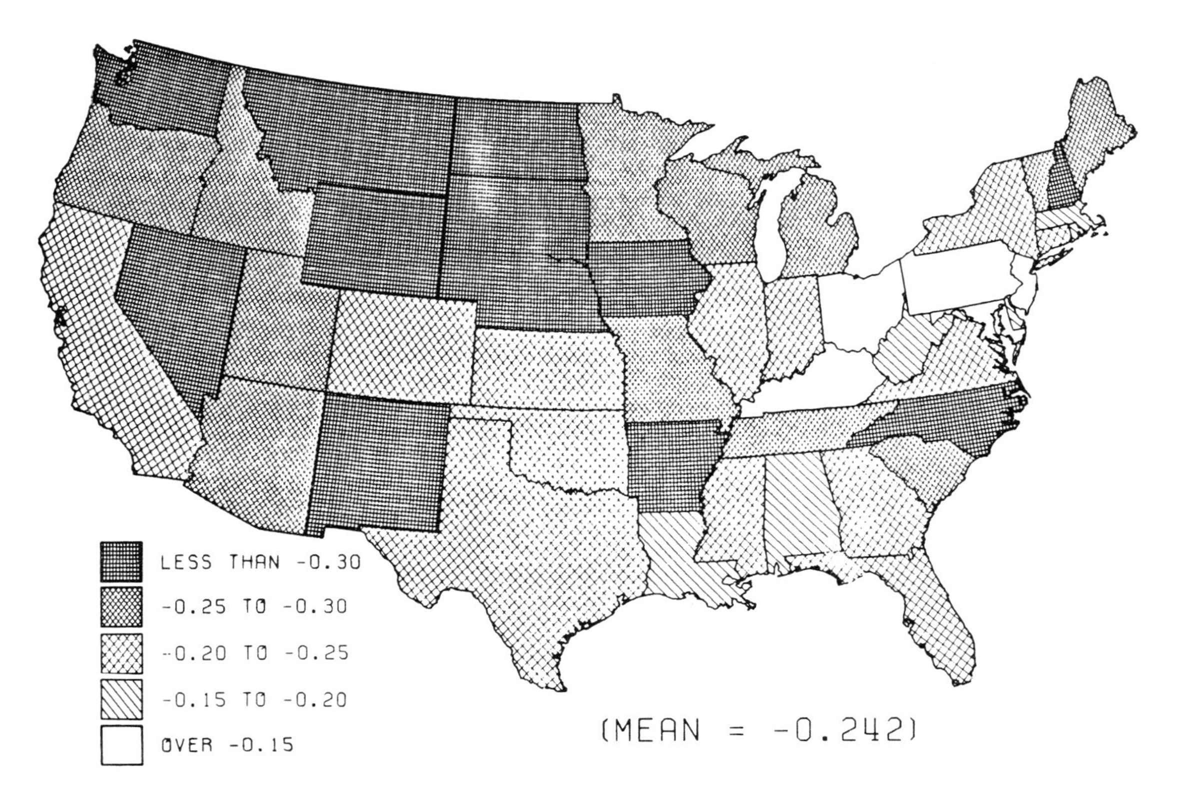

Figure 1. Estimated Gasoline Price Elasticities, Jan/1975--July/1980.

TABLE 1. ESTIMATED PARAMETER VALUES

STATE	PRICE	INCOME	MAY	JUNE	JULY	NOISE MODELS	λ^2/d.f.
Alabama	-0.167	0.627	-0.013*	-0.056	-0.082	$\frac{a_t}{(1 - B^{12})}$	36.2/30
Alaska	-0.173*	0.163*	0.015	0.127	-0.032	$\frac{a_t}{(1 - B^{12})}$	38.6/30
Arizona	-0.297	0.319	-0.022*	-0.051*	-0.057*	$\frac{(1 - 0.455B - 0.472B^4)a_t}{(1 - B^{12})}$	21.4/28
Arkansas	-0.310	0.328*	0.035*	-0.121*	-0.052*	$\frac{(1 - 0.433B)a_t}{(1 - B^{12})}$	32.4/29
California	-0.235	0.543	-0.028*	-0.003*	-0.052	$\frac{(1 + 0.491B^2)a_t}{(1 - B^{12})}$	38.4/29
Colorado	-0.214	0.395	0.024*	-0.009*	-0.047	$\frac{(1 - 0.595B)a_t}{(1 - 0.837B^1)(1 - B^{12})}$	30.1/28
Connecticut	-0.220	0.525	-0.048	-0.035	-0.033**	$\frac{(1 + 0.502B^3)a_t}{(1 - B^{12})}$	20.5/29
Delaware†	-0.156	0.084*	0.135**	-0.071*	-0.024*	$\frac{a_t}{(1 - B^{12})}$	28.3/30
District of Columbia	-0.228	-0.114*	-0.012*	-0.085*	-0.108*	$\frac{(1 - 0.824B^{12})a_t}{(1 - B^{12})}$	17.9/29

TABLE 1. (Continued)

STATE	PRICE	INCOME	MAY	JUNE	JULY	NOISE MODELS	λ^2/d.f.
Florida	-0.204	0.337	-0.037*	-0.027*	-0.039*	$\frac{(1 - 0.336B^2 - 0.367B^7)a_t}{(1 - B^{12})}$	38.2/28
Georgia	-0.238	0.451	-0.010*	-0.018*	-0.027*	$\frac{a_t}{(1 - B^{12})}$	32.9/29
Hawaii	-0.004*	0.109*	0.006*	-0.024*	-0.059*	$\frac{(1 - 0.307B^{12})a_t}{(1 - B^{12})}$	28.6/29
Idaho	-0.257	0.378**	-0.082*	0.083*	-0.085*	$\frac{(1 -).426B^{12})a_t}{(1 + 0.625B)(1 - B^{12})}$	26.4/28
Illinois	-0.248	0.506	0.030*	-0.016*	-0.045**	$\frac{a_t}{(1 - B^{12})}$	43.7/30
Indiana	-0.238	0.638	-0.003*	-0.058*	-0.063*	$\frac{a_t}{(1 - B^{12})}$	35.2/30
Iowa	-0.377	-0.123*	-0.024*	-0.023*	-0.017*	$\frac{a_t}{(1 - B^{12})}$	32.8/30
Kansas	-0.208	0.200*	-0.055*	-0.028*	-0.043*	$\frac{a_t}{(1 + 0.590B^{12})(1 - B^{12})}$	19.5/29
Kentucky	-0.123*	0.861	-0.080	-0.053*	-0.079	$\frac{a_t}{(1 - B^{12})}$	40.6/30

TABLE 1. (Continued)

STATE	PRICE	INCOME	MAY	JUNE	JULY	NOISE MODELS	λ^2/d.f.
Louisiana	-0.196	0.848	-0.022*	-0.024*	-0.019*	$\frac{(1 - 0.476B - 0.487B^{12})a_t}{(1 - B^{12})}$	13.1/28
Maine	-0.288	0.547	0.0*	-0.053	-0.078	$\frac{(1 + 0.419B^{3})a_t}{(1 - B^{12})}$	29.8/29
Maryland	-0.148	0.663	-0.040*	-0.093	-0.110	$\frac{a_t}{(1 - B^{12})}$	28.7/30
Massachusetts	-0.180	0.944	-0.024*	-0.025*	-0.029*	$\frac{(1 - 0.579B^{12})a_t}{(1 - B^{12})}$	34.6/29
Michigan	-0.297	0.482	-0.041*	-0.042*	-0.073*	$\frac{(1 - 0.784B^{12})a_t}{(1 - B^{12})}$	21.6/29
Minnesota	-0.289	0.085*	0.007*	-0.011*	-0.052*	$\frac{(1 - 0.427B + 0.411B^{10})a_t}{(1 - B^{12})}$	26.2/28
Mississippi	-0.240	0.511	-0.037*	-0.078**	0.188	$\frac{(1 - 0.336B^{2} + 0.640B^{3})a_t}{(1 - B^{12})}$	29.2/28
Missouri	-0.276	0.388	-0.018*	-0.049*	-0.025*	$\frac{a_t}{(1 - B^{12})}$	24.9/30
Montana	-0.357	-0.082*	-0.007*	0.022*	-0.029*	$\frac{(1 - 0.418B - 0.331B^{2})a_t}{(1 - B^{12})}$	31.7/28

TABLE 1. (Continued)

STATE	PRICE	INCOME	MAY	JUNE	JULY	NOISE MODELS	λ^2/d.f.
Nebraska	-0.355	-0.151*	-0.031*	0.024*	-0.016*	$\frac{a_t}{(1 - B^{12})}$	21.1/30
Nevada	-0.343	0.094*	-0.016*	-0.022*	-0.027*	$\frac{(1 + 0.436B^2)a_t}{(1 - B^{12})}$	21.0/29
New Hampshire	-0.330	0.575	-0.015*	-0.071	-0.054	$\frac{(1 + 0.798B^3)a_t}{(1 - B^{12})}$	35.6/29
New Jersey	-0.142	0.843	-0.052*	-0.104	-0.055*	$\frac{(1 - 0.473B^{12})a_t}{(1 - B^{12})}$	38.1/29
New Mexico	-0.320	0.623	-0.077*	-0.046*	-0.113*	$\frac{(1 - 0.678B^{12})a_t}{(1 + 0.409B)(1 - B^{12})}$	17.1/28
New York	-0.211	0.435*	-0.029*	-0.050*	-0.049*	$\frac{(1 - 0.308B)a_t}{(1 - B^{12})}$	25.1/29
North Carolina	-0.326	0.493	-0.059	-0.037*	0.047*	$\frac{a_t}{(1 - B^{12})}$	29.2/30
North Dakota	-0.326	0.229*	0.014*	0.032*	-0.040	$\frac{a_t}{(1 - B^{12})}$	25.2/30

TABLE 1. (Continued)

STATE	PRICE	INCOME	MAY	JUNE	JULY	NOISE MODELS	λ^2/d.f.
Ohio	-0.138	0.711	-0.016*	-0.069*	-0.020*	$\frac{(1 - 0.657B^{12})a_t}{(1 - B^{12})}$	23.2/29
Oklahoma	-0.233	0.250	-0.014*	-0.097*	-0.036*	$\frac{(1 - 0.490B - 0.870B^{12})a_t}{(1 - B^{12})}$	29.2/28
Oregon	-0.265	0.098*	-0.026*	0.001*	-0.067	$\frac{a_t}{(1 - B^{12})}$	27.1/30
Pennsylvania	-0.036*	1.710**	-0.090*	-0.122*	0.130*	$\frac{(1 - 0.284B)a_t}{(1 - B^{12})}$	14.1/29
Rhode Island	-0.169	0.571	-0.003*	-0.004*	-0.101	$\frac{(1 - 0.472B^{12})a_t}{(1 - B^{12})}$	18.5/29
South Carolina	-0.274	0.581	-0.004*	-0.057*	-0.032*	$\frac{(1 - 0.586B^{12})a_t}{(1 - B^{12})}$	19.6/29
South Dakota	-0.334	0.067*	0.111*	0.019*	-0.023*	$\frac{(1 - 0.381B^{4})a_t}{(1 - B^{12})}$	19.8/29
Tennessee	-0.289	0.505	-0.010*	-0.057*	-0.043*	$\frac{(1 - 0.480B^{2} - 0.528B^{12})a_t}{(1 - B^{12})}$	25.8/28
Texas	-0.226	0.408**	-0.060*	0.006*	0.019*	$\frac{a_t}{(1 - B^{12})}$	41.0/30

TABLE 1. (Continued)

STATE	PRICE	INCOME	MAY	JUNE	JULY	NOISE MODELS	λ^2/d.f.
Utah	-0.251	0.339	-0.084	-0.046*	-0.041*	$\frac{(1 - 0.329B^6)a_t}{(1 - B^{12})}$	29.3/29
Vermont	-0.271	0.398*	-0.026*	-0.035*	-0.032*	$\frac{a_t}{(1 - B^{12})}$	31.5/30
Virginia	-0.230	0.838	-0.022*	-0.075	-0.041	$\frac{a_t}{(1 - B^{12})}$	37.3/30
Washington	-0.314	0.333	-0.003*	-0.017*	-0.012*	$\frac{(1 + 0.341B^2)a_t}{(1 - B^{12})}$	34.6/29
West Virginia	-0.171	0.656	-0.048*	-0.016*	-0.041*	$\frac{a_t}{(1 - B^{12})}$	38.4/30
Wisconsin	-0.264	0.504	-0.030*	-0.021*	-0.039*	$\frac{a_t}{(1 - B^{12})}$	35.8/30
Wyoming	-0.348	0.150*	-0.018*	0.007*	-0.056*	$\frac{a_t}{(1 - B^{12})}$	25.3/30

* Indicates estimated coefficient not significant even at 0.1 level.

† Includes 64 data points only.

** Indicates estimated coefficient significant at 0.1 but not 0.05 level.

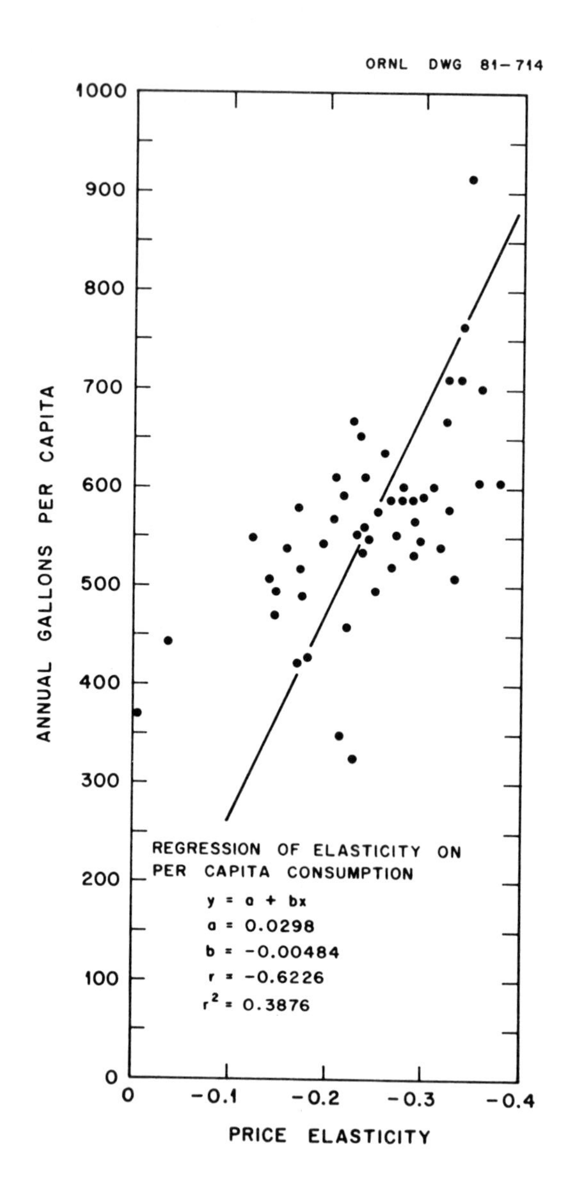

Figure 2. Association of Price Elasticity and Consumption Per Capita.

It would be a mistake, of course, to push this simple analysis too far. Yet response to a fuel shortage is critically sensitive to whether price elasticities increase, decrease, or remain constant as prices rise. The modeling and analysis in this report assume constant elasticities which are consistent with the literature on the subject (e.g., Danielson and Agarwal, 1976; Greene, 1981). The simple analysis just presented suggests that if elasticities are not constant they decrease as prices rise. The issue is important since a shortage of 20% would cause a 200% increase in prices at an elasticity of -0.2 and a 340% increase at an elasticity of -0.15. In the analysis of supply shortage management strategies, constant elasticities will be assumed. If elasticity varies directly with the level of consumption, our results will be conservative.

The estimates presented in Table 1 imply large variabilities in the responses of different regions to energy supply shortages. In the northern mountain and plains states where price elasticities less than -0.3 are common, a 20% shortage might be expected to produce price increases in the vicinity of 100%. In the northeast the potential for 350% price increases exists. Assuming the regions were isolated (i.e., no arbitrage was allowed) the price of a gallon of gas in the northeast might be $5.60 while the same commodity in Montana, Wyoming, or the Dakotas would sell for only $2.50. Obviously great financial incentives to redistribute supply would exist. As an illustration of the kind of supply redistribution which might take place, assume that a national price of $3.20 was established (this is equivalent to a national price elasticity of -0.238 and a 20% shortage). At $3.20 instead of $1.25, Iowa, with a price elasticity of -0.377, would decrease its consumption by 30%. On the other hand, Ohio, with an elasticity of only -0.138 would cut back a mere 12%. If both started with 80% of their normal supply, Ohians would want to import an additional 8% while Iowans would want to export 10% of their normal consumption.

The first question that arises is whether the petroleum and petroleum products transportation system are capable of redistributing supplies in the manner that would be required. This is far from being given, even in the case of fuel shortage since water and pipeline transport systems, which are the modes used to transport the bulk of petroleum and petroleum products, are inflexible with respect to where they can deliver and, to some extent, even the direction of movement. Unfortunately this question is beyond the scope of this paper.

A second, and possibly equally important question, is whether the energy supply emergency management strategy adopted will facilitate or impede the redistribution of available supplies. Various rationing, taxation, or market schemes all create differing incentives to redistribute supply. In the following section a typology of shortage management schemes is developed. Each approach is then analyzed with respect to its economic efficiency and income redistributive characteristics as well as its effect on incentives to efficiently redistribute available supply.

4. STRATEGIES FOR MANAGING A GASOLINE SUPPLY EMERGENCY

Several strategies for allocating gasoline during a supply emergency have been proposed and analyzed (see, e.g., Difiglio, 1980, upon which this section draws heavily). Those which work through (or in opposition to) the economic market in gasoline can be broadly grouped into those that employ price controls and those which allow prices to be determined in the market. The simplest possible type of each takes no other action than to either freeze prices or allow them to float. Within the category of price control strategies, types of plans are distinguished by the method used to redistribute income or, by fiat, restrain demand. Within the category of uncontrolled price strategies, plans may be distinguished according to the type of taxing mechanism used to distribute the huge increase in profits caused by the shortage. Table 2 lists the strategy types whose regional and income redistribute effects will be considered.

TABLE 2. A TYPOLOGY OF DEMAND SIDE SHORTAGE STRATEGIES

I. Price Control	II. Uncontrolled prices
1. Fixed price only	1. Market prices only
2. Fixed price with demand restraining regulations	2. Floating price with fixed tax
3. Fixed price with fixed tax	3. Floating price with prorated windfall profits tax
4. Fixed price with commodity currency (rationing) (a) nonnegotiable coupons (b) negotiable coupons	4. Floating price with total windfall profits tax

Price control strategies are probably most often thought of when emergency planning strategies are considered. The Emergency Energy Conservation Act of 1979 (U.S. Congress, 1979) uses a combination of 1.2 and 1.4.b to deal with shortages of differing sizes. The category includes price control and rationing by queuing which we have seen in the 1973-74 and 1979 shortages, as well as gasoline rationing which has not been tried since WWII. In general, Type I strategies tend to produce market shortages with the possible exception of negotiable coupon rationing. Type II strategies, on the other hand, use prices to regulate the market and tend to result in market clearing conditions. The apparent advantage of Type II strategies is that they do not allow windfall profits and thus avoid the social conflicts which would probably attend large transfers of income from consumers to the holders of scarce oil resources. As has been pointed out above these transfers would be large indeed.

The properties of each strategy in terms of economic efficiency, income redistribution, and effect on the regional redistribution of resources can be illustrated simply and effectively by means of two dimensional graphical representations of supply and demand.

5. PRICE CONTROL STRATEGIES

Under simple price controls, prices are held constant by law at their pre-shortage level. For simplicity, we assume this to be $1 per gallon. The shortage is assumed to change the supply curve from essentially horizontal at SS' to strictly vertical S_2S_2' (Figure 3). This assumption rules out the purchase of additional supply on world spot markets because international agreements in which the United States participates effectively prohibit such an action (e.g., Alm, 1981, p. 1383).* In brief, we assume an absolute gasoline supply shortage of 20%.

With supply restricted to Q_2 the market clearing price would be P_2 = $3.05 per gallon (assuming a national price elasticity of -0.2). This is the amount consumers would be willing to pay, but since they are prevented (or protected) from doing so, the available supply will be rationed by waiting on line. This has three undesirable effects. Difiglio (1980) has speculated rather conservatively that a 30% shortage might result in mean waiting times of two hours and that this would result in annual deadweight economic losses of 60 billion dollars in consumers' time (valued at $4 per hour). In addition, Difiglio estimates that 16 billion dollars worth of fuel would be wasted by engines idling while waiting (325,000 bbls per day valued at $3.20 gallon). Another way to estimate the total deadweight loss is to assume that consumers would, in effect, pay P_2 = $3.05 gallon for fuel in time, wasted fuel, and aggravation. Thus the deadweight loss would be $Q_2 (P_2 - P_1) = \$164 \times 10^9$ if $Q_2 = 80 \times 10^9$ gallon (approximately 80 percent of the present level of consumption). In either case the economic loss is staggering.

The third disadvantage of the price control case is that it provides no incentive to redistribute available supplies from states with high price elasticities to states with inelastic demand. Thus, regional variations in shortages are virtually assured. Since fuel cannot be sold at a higher price where it is in greatest demand and since all available supplies could be sold in any given location there would be no reason for suppliers to shift supplies from states with short queues to states with longer ones. Extreme regional inequities would inevitably result. Consider that the queuing price in a state with a price elasticity of -0.3 would be $2.10 - $1.00 = $1.10 while it would be $4.40 - $1.00 = $3.40 where the price elasticity was only -0.15. Low elasticity regions would pay an extra $2.00 gallon in waiting time, aggravation, and wasted fuel.

The effect of demand restraining regulations is to transfer some of the deadweight loss due to waiting time and wasted fuel to various other nuisances caused by the regulations. For example, reducing speed limits substitutes increased travel time for waiting time and fuel. Purchase control schemes presumably substitute inconvenience for waiting time and fuel. In essence, the deadweight loss should be unchanged. The only way to reduce deadweight loss given fixed prices is to substitute information and intelligence. For example, informing

---

* Under an agreement with the International Energy Agency the United States could lose 75% of its imports (Alm, 1981, p. 1381) which in 1980 would have been 30% of total supply. Difiglio (1980, p. 1) considers a 40% shortage possible.

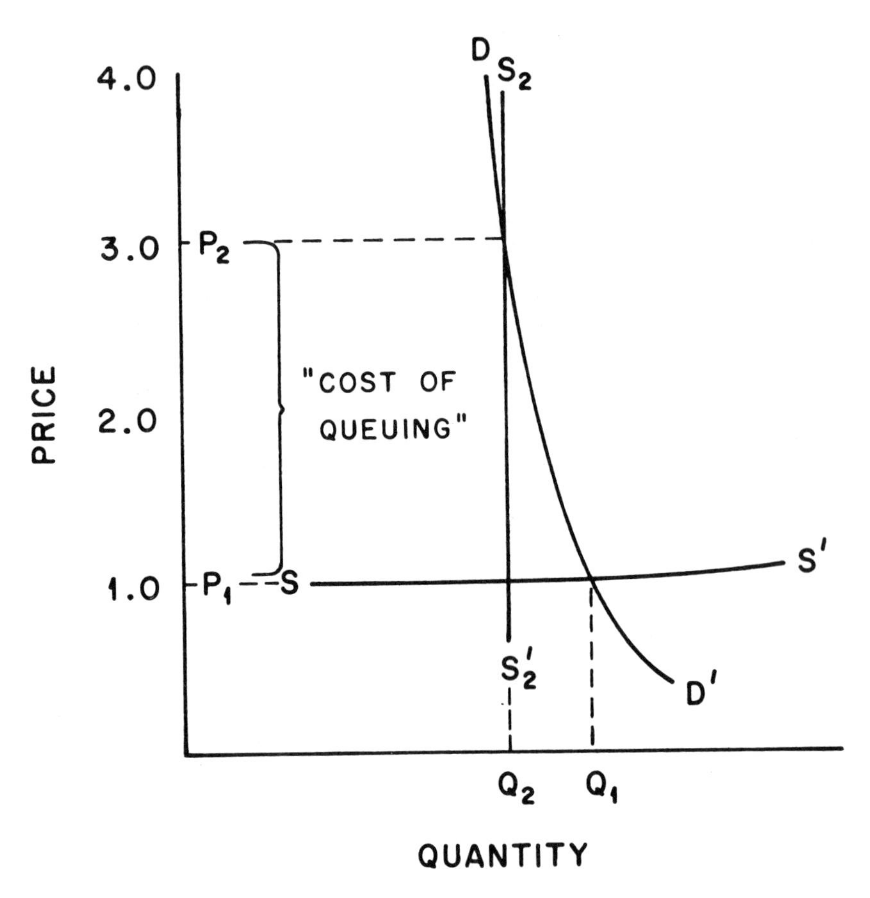

Figure 3. Simple Price Control.

consumers that they can save fuel by increasing tire pressures has essentially no cost but can improve fuel economy several percent. Likewise, more intelligent driving strategies may also save substantial amounts of fuel at little or no cost (Greene, 1981). Such actions, which effectively increase the price elasticity of demand, are very likely the most desirable of all demand-side emergency strategies.

Of course, prices can be fixed at any level. An interesting case is when a fixed tax is added which is less than the difference between the market clearing price and the original price. Seen from the supplier's point of view, this has the effect of shifting the demand curve downward. However, it can also be represented as a simple increase in the controlled price with no change in the demand curve (Figure 4). In this case we assume a simple $1 gallon tax is imposed. The only change over the simple price control case is that waiting time costs and fuel losses due to idling are reduced. Although deadweight economic losses are reduced, the total cost of fuel to consumers remains the same except that some of that has been collected by the government and can be redistributed back to consumers. Difiglio (1980) has suggested how this could be done in a fairly simple and reasonably equitable manner through the Internal Revenue Service. It is crucial, of course, that rebates to consumers are not a function of actual purchases.

Just as in the simple price control case, there is no incentive to geographicaly redistribute supplies unless the tax is so high that it exceeds the regional market clearing price. If that were to happen a regional surplus would result. This would create a speculative situation for suppliers. They could stockpile supplies under the assumption that prices would increase in the future, they could cut regional prices (probably selling at a loss), or they could shift supplies to regions where there were shortages at the taxed price. As we have noted above, such a redistribution may be much more difficult than it sounds. Rigidities in the petroleum and product distribution system might make the transport costs prohibitive. The required redistribution might not even be feasible. Furthermore, fixed prices would not allow suppliers to recoup any of these additional costs. Under the circumstances, the optimal decision for suppliers might well be to cut back production for regions in surplus and not supply additional fuel to deficit regions. Unfortunately, this important question is beyond our scope.

In sum, the fixed-price fixed-tax strategy will also lead to regional inequities. The nature of these will depend on how close the taxed price comes to the national market clearing price. If it is far below the market clearing price then shortages will exist in all regions. The size of the shortage will vary with the regional price elasticities. In regions where demand is price elastic there will be very short queues. Where demand is inelastic mean waiting times may be two hours or more. If the price is close to the national market clearing price, price elastic regions will be in surplus and inelastic regions in deficit. What will happen then is largely unknown.

The final price control strategy is coupon rationing. Coupon rationing creates a commodity-specific currency which may be either negotiable or nonnegotiable. Under nonnegotiable coupon rationing the

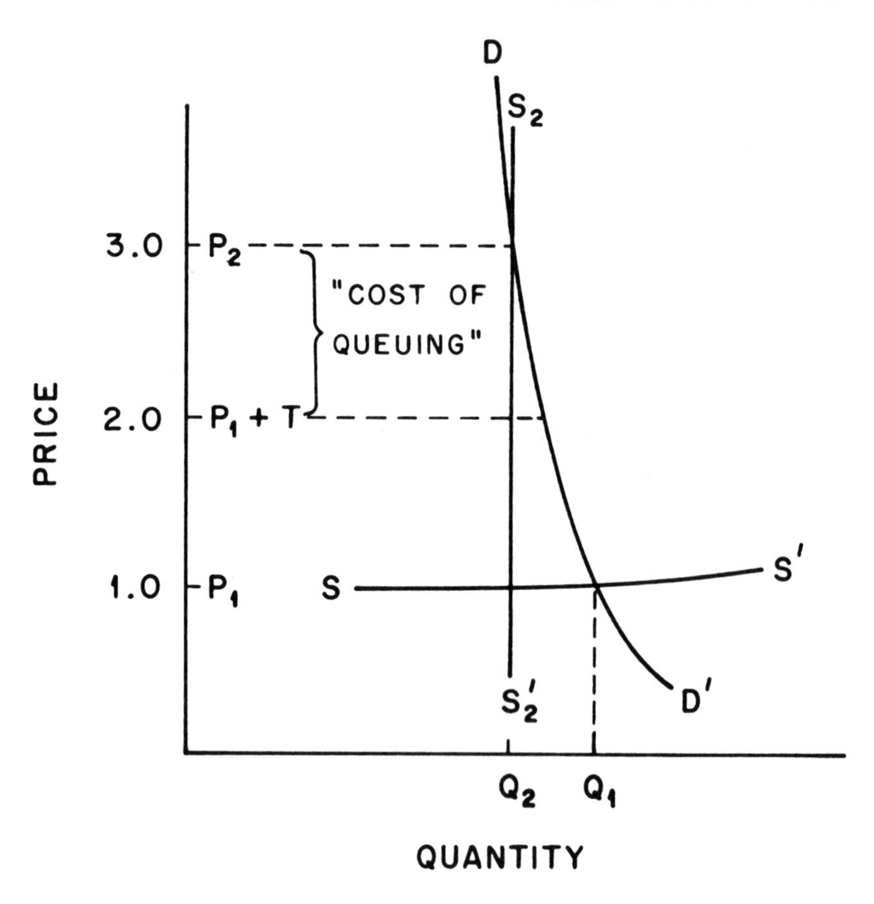

Figure 4. Fixed Price and Fixed Tax.

coupon grants the person to whom it is issued a legal right to purchase gasoline which cannot be sold (although it may be transferable _gratis_). Under the nonnegotiable coupon system supply and demand are equilibrated by issuing a quantity of coupons equal to the available supply. Thus supply and demand exist only at the point A in Figure 5. With or without a black market there is essentially no deadweight economic loss due to waiting on gasoline lines. There is, however, a consumer surplus loss because consumers are not allowed to engage in arbitrage for coupons. For example, person A might prefer to pay person B $1 for a coupon while person B would prefer $1 to one coupon. If the exchange were allowed both would be better off. Since no exchanges are allowed consumers are confined to lower levels of satisfaction than they could achieve through exchange. An estimate of this quantity is beyond our scope though it is clear that even moderate differences in price elasticities across individuals would place its value easily in the tens of billions of dollars on an annual basis. The administrative costs of rationing, however, would be considerable since issuing coupons amounts to issuing a new currency similar to postage stamps. The Department of Energy has estimated that gasoline rationing would cost between 2 and 4 billion dollars per year to administer and would involve 40 to 50 thousand government employees (U.S. DOE, 1980). While this is small, relative to potential deadweight losses to consumers, it is still a substantial amount. Probably more important is the fact that a black market would almost surely develop. The costs of a black market are partly social costs such as the loss of respect for legal authority although there would certainly be some enforcement costs as well. The black market, however, would tend to reduce consumer surplus loss due to the inherent economic inefficiency of coupon rationing.

A way to avoid the social problems of black markets and the economic inefficiency of nonnegotiable coupon rationing is to permit the buying and selling of coupons. The effect of this is to allow the price of gasoline plus coupons to reach the market clearing level of P_2 (Figure 5). In the 20% shortfall, -0.2 price elasticity example a single coupon would sell at $P_2 = P_1 = \$2.05$, more than twice the original price of gasoline. If 80×10^9 gallons are sold on an annual basis, then the total value of all coupons distributed in a single year is $\$164 \times 10^9$. The distribution of coupons in effect constitutes a distribution of income in the form of rights to gasoline purchase. This distribution can be handled in almost any way desired. Difiglio (1980, p. 34) has shown that a distribution based on number of vehicles owned results in a reasonably progressive income redistribution.

It is clear that regional variations in price elasticities would result initially in regional variations in coupon prices. Coupons must be geographically mobile so that inelastic demand regions can purchase from elastic demand regions resulting in a national coupon price. To facilitate this, some sort of national banking system or exchange markets would be necessary. To the extent that the original distribution of coupons differed from the equilibrium demand under white market rationing there would be interregional income transfers.

In an example given above it was shown that at a national price of

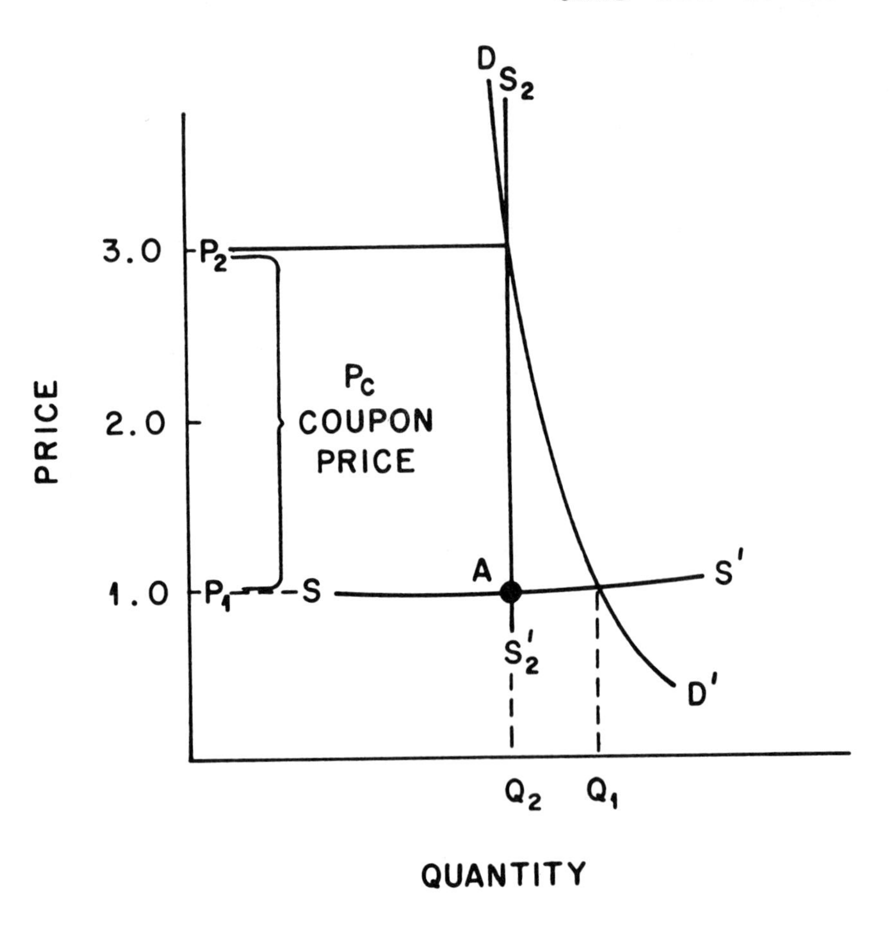

Figure 5. Coupon Rationing.

$3.20 per gallon (assuming a price elasticity of -0.238 and an original price of $1.25) Ohians would want to import an additional 8% while Iowans would prefer to sell an additional 10% of their normal consumption. Total imports and exports for these states would be approximately 425 and 175 million gallons, respectively, or 1.36 and 0.56 billion dollars on an annual basis. All other states would have smaller relative transfers since these states represent the two extremes of price elasticity estimates. Finally, as has been pointed out before, there is no guarantee that fuel supplies will be mobile even if coupons are. Under coupon rationing there is no premium offered to suppliers for redistributing supplies. Suppliers receive only the initial, regulated price. Exactly what would take place under these circumstances is once again uncertain. Suppliers might stockpile against future sales or might choose to incur the additional transport costs even if it meant selling at a loss. The latter choice seems unlikely since if suppliers did stockpile in surplus regions the cost of coupons in deficit regions would decline because even coupon holders would have to wait in line to purchase fuel. To some extent coupons would flow back to surplus regions enabling consumers there to buy up some of the excess supply. Just how the difference $P_2 - P_1$ would be divided between coupon price and queuing costs would depend on the relationship between excess demand (the difference between coupons available and supplies available) and average queuing times. This division, in turn, would be simultaneously determined with the quantity of coupons imported. Clearly this issue deserves further study before the white market, coupon rationing system is chosen.

In conclusion, it is apparent that any fixed price rationing scheme adopted will result in some economic loss. These losses can be staggering, as in the case of simple price control where losses of $75 to $165 x 10^9 are likely on an annual basis for a 20% shortage. Adding a fixed tax with rebate can reduce deadweight losses due to queuing considerably but may create an inefficient regional allocation of available supplies. It may also create regional surpluses and deficits without providing a financial incentive for suppliers to redistribute available supplies. Non-negotiable coupon rationing eliminates queuing losses, but produces an inefficient allocation and would almost certainly create an illegal black market. Negotiable coupon rationing would at first appear to solve all these problems. Even income transfer effects can be controlled reasonably well via the distribution of rationing coupons. Administrative costs, however, would not be inconsiderable. More importantly, the existence of regional variations in price elasticities makes it likely that gasoline lines and their deadweight economic loss plus inefficient resource allocation would exist to a presently unknown extent. The key issue is whether supplies could and would be redistributed to match new equilibrium pattern of regional demand. The question is both of feasibility given the present commodity transportation infrastructure, and of producer behavior given no financial inducement to incur additional transport costs.

6. UNCONTROLLED PRICE STRATEGIES

Uncontrolled price strategies are inherently simpler than price control strategies since they rely on existing market mechanisms to achieve market clearing price. Because of this they tend to result in an economically efficient allocation of resources without shortages and the attendant gasoline queues. However, income transfers and regional redistributions of supplies are still potential problems.

The simple uncontrolled price case is a null plan indeed. When a shortage occurs the market price will simply shift from P_1 to P_2 (Figure 6). The most dramatic result is a massive transfer of income in the amount of $(P_2 - P_1)\ Q_2$ from gasoline purchasers to the owners of the suddenly scarce resource. The amount of the transfer is nothing short of mind boggling. It was computed in Section 2 of this paper at about 0.57 billion dollars per day (1980 data). On an annual basis this is equal to eighty percent of all government transfer payments in 1980. The importance of this issue can hardly be over-emphasized. The null plan would result in such an enormous windfall for owners of gasoline resources at the expense of consumers that it is hard to see how the problem could be ignored.

A less obvious question is that of regional supply redistribution. Greene and Chen (1981) have presented evidence based on 1975-1980 data that the national price elasticity is about -0.2. If the initial price is $1.25 then the national price given a 20% shortage would be

$$\frac{q_{t+1}}{q_t}^{-0.2} \cdot p_t = p_{t+1}$$

$$(.8)^{-5.0} \ . \ 1.25 = \$3.81$$

Using the state price elasticities presented in Table 1, we can compute the percent of initial consumption which would be demanded in each state. The results for the ten most extreme states are presented in Table 3. While the redistributions required are quite large, they are only two or three times as large as changes which have occurred in the past over the period of a year. The question, however, is not whether such a redistribution could be gradually accommodated under normal circumstances, but rather whether such changes in flows could be quickly carried out when normal sources of supply have been severely disrupted. The answer is unfortunately unknown.

The alternative to simple uncontrolled prices is to tax away some of the difference between original and shortage market clearing prices by means of a fixed or pro rata tax. Proceeds from this tax could then be returned to consumers according to any desired formula so long as it does not depend on actual quantities consumed. Suppliers would also receive additional profits to the extent that $P_2 - \text{Tax} > P_1$. This payment to producers would buy an efficient market allocation and would serve as a direct monetary incentive to regionally redistribute supply.

The fixed tax case is represented in Figure 6. A $1.55/gal. tax is assumed. From the suppliers' point of view the tax shifts the demand curve downward (to $D_T D_{T'}$) so that the market equilibrium price is just $1.50 per gallon, or 50% higher than the original price. Once again the national level analysis masks what is happening to regional markets. Assume again that each state begins with 80% of its normal supply. Ohio wishes to buy more gasoline and will pay $5.04/gallon, of which $5.04 - $1.50 = $3.54 would go to the supplier. On the other hand, Iowa would prefer to sell some of its available supply. In fact,

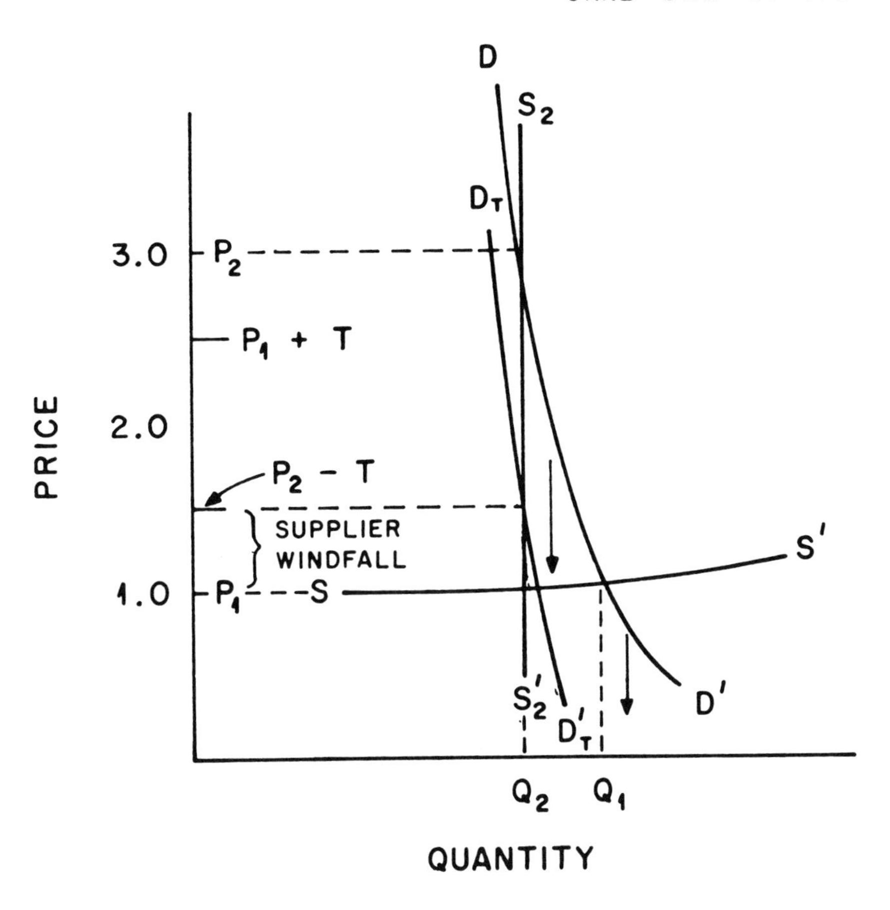

Figure 6. Uncontrolled Prices and a Fixed Tax.

TABLE 3
PRICE ELASTICITY INDUCED IMPORTS AND EXPORTS FROM THE TEN MOST EXTREME STATES

State	Price elasticity	% of 1980 demand given $3.81 gal. price	80% of 1979 demand* (10^6 gals)	Import (-) Export (+) (10^9 gallons)
Iowa	-0.377	66	1375	+240
Montana	-0.357	67	400	+ 65
Nebraska	-0.355	67	726	+118
Wyoming	-0.348	68	305	+ 46
Nevada	-0.343	68	401	+ 60
Ohio	-0.138	86	4285	-268
New Jersey	-0.142	85	2716	-170
Maryland	-0.148	85	1561	- 98
Delaware	-0.156	84	245	- 12
Alabama	-0.167	83	1675	- 63

* Final 1980 motor gasoline consumption figures were not available at the time of writing. Source for 1979 motor gasoline consumption: U.S. Dept. of Transportation, Federal Highway Administration, Highway statistics 1979, Table MF-33GA, U.S. Government Printing Office, Washington, D.C.

with a 20% shortage Iowans are only willing to pay $1.80/gallon so that suppliers would lose money by selling gas there. In fact, the incentive to shift supplies from Iowa to Ohio is a whopping $3.24/gallon price difference. This kind of inducement essentially assures an efficient redistribution of resources if it is physically possible. A *pro rata* or percentate tax would have essentially the same effect except that extreme differences across states would not result in such large incentives for suppliers to redistibute supplies.

Uncontrolled price rationing tends to produce efficient, market clearing solutions, but, in the absence of government taxation, would also produce a tremendous redistribution of income from consumers to owners of gasoline resources. Simple tax schemes can be devised, however, which retain the efficient market clearing properties, but return much of the windfall profits to consumers. Furthermore, taxing windfall profits leaves substantial financial incentives for suppliers to redistribute supplies to alleviate regional shortage.

7. CONCLUSION

Any gasoline supply shortage management strategy must address two major issues: 1) the social cost of petroleum shortage in terms of time lost in gasoline lines, fuel wasted while waiting on line and administrative costs, and 2) the potential redistribution of income between consumers and owners of gasoline resources. Social costs from a 20% shortage could be between $75 and $165 billion on an annual basis. Income redistribution could well exceed $200 billion per year. Price control strategies prevent large income transfers, but tend to result in economically inefficient allocations of supply and, unless rationing is adopted, would waste tens of billions of dollars of consumers' time in gasoline lines. Unless regional supply redistribution can be achieved at virtually no cost, rationing schemes, whether with negotiable or nonnegotiable coupons, will likely lead to regional shortages and surpluses since they offer no financial inducement to producers to redistribute supplies.

Uncontrolled price strategies eliminate problems of deadweight economic cost (except for administrative costs). On the other hand, unless windfall profits are taxed effectively, they allow phenomenal transfers of income from consumers to gasoline resource owners. Equitable taxing schemes can be devised, however, which not only restore most of consumers' income but also retain substantial incentives for suppliers to equilibriate regional gasoline markets by redistributing supplies. Unfortunately, doubt still remains about the ability of the petroleum and product transport system to actually carry out the necessary transshipment of supplies during an interruption of normal supplies. This issue must be addressed if the United States is to be adequately prepared to deal with a petroleum supply emergency.

BIBLIOGRAPHY

Alm, L., 'Energy Supply Interruptions and National Security,' Science, Vol. 221, No. 4489, pp. 1379-1385, March 27,1981.

Box, G.E.P. and G. M. Jenkins, Time Series Analysis, Forecasting and control, Holden-Day, Inc., San Francisco. (1970).

Danielson, A. L. and M. L. Agarwal, 'A Functional Form Analysis of the Demand for Refined Petroleum Products,' Proceedings of the Third Annual UMR-MEC Conference on Energy, University of Missouri-Rolla, October 12-14, 1976.

Digiflio, C., 'Economic Allocation of Gasoline Shortages,' presented at the National Energy Users Conference for Transportation, sponsored by the Transportation Research Board, April 13-16, San Antonio, Texas. (1980).

Greene, D. L., 'Regional Demand for Gasoline: Comment,' Journal of Regional Science, Vol. 20, No. 1, pp. 103-109. (1980a).

Greene, D. L., 'The Spatial Dimension of Gasoline Demand: An Econometric Analysis of State Gasoline Use,' Geog. Survey, Vol. 9, No. 2, pp. 19-28, April, 1980b.

Greene, D. L., 'A State Level Stock System Model ofGasoline Demand,' forthcoming, Transportation Research Record, Transportation Research Board, Washington, D.C. (1981a).

Greene, D. L., 'Testimony Before the Interior Appropriations Subcommittee of the Appropriations Committee of the United States House of Representatives,' April 8, Washington, D.C. (1981b).

Johnson, L. W., 'Regional Demand for Gasoline: Comment,' Journal of Regional Science., Vol. 20, No. 1, pp. 99-100. (1980).

Kraft, J. and M. Rodekohr, 'Regional Demand for Gasoline A Temporal Cross-Section Specification,' Journal of Regional Science, Vol. 18, No. 1, pp. 45-55. (1978).

Kraft, J. and M. Rodekohr, 'Regional Demand for Gasoline: A Reply to some Reconsiderations,' Journal of Regional Science, Vol. 20, No. 1, pp. 111-114. (1980).

Mehta, J.S., G.V.L. Narasimham and P.A.V.B. Swamy, 'Estimation of a Dynamic Demand Function for Gasoline,' Journal of Econometrics, Vol. 7, pp. 263-279. (1978).

Osleeb, J.P., 'An Evaluation of the Strategic Petroleum Reserve Program of the United States Department of Energy,' The Professional Geographer, Vol. 31, No. 4, pp. 393-399. (1979).

Platt's Oil Price Handbook and Oilmanac, eds. 52-56 'Service Station Prices, Gasoline (Including Taxes),' McGraw-Hill, Inc., New York. (1975-1979).

U.S. Department of Commerce, Bureau of Economic Analysis, Survey of Current Business, Vols. 56-61 'Commodity Prices,' 'General Business Indicators - Monthly Series, Federal Government Receipts and Expenditures,' U.S. Government Printing Office, Washington, D.C. (1976-1981).

U.S. Department of Commerce, Bureau of the Census, Statistical Abstract of the United States, 1977, Table No. 10, 'Population-States: 1960 to 1976,' p. 11, U.S. Government Printing Office, Washington, D.C.

U.S. Department of Commerce, Bureau of the Census, '1980 Census Counts of the Population of States, By Region and Division,' final apportionment counts, unpublished table, January. (1981).

U.S. Departmnent of Energy,(1980). Monthly Energy Review, DOE/EIA-0035(80/11), November, U.S. Government Printing Office, Washington, D.C. (1980).

U.S. Department of Energy, 'Managing Motor Fuel Shortages,' NEP-11 Issue Paper, March 6, Washington, D.C. (1980).

U.S. Senate, U.S. Senate Conference Report 96-366, 'Emergency Energy Conservation Act of 1979,' p. 11. (1979).

U.S. Department of Transportation, Federal Highway Administration, 'Monthly Motor Gasoline Use,' Table MF-33G, unpublished, Washington, D.C.

'Monthly Personal Income Figures by States,' courtesy of Business Week, McGraw-Hill, New York.